PRINT READING
FOR MACHINISTS

SIXTH EDITION | DAVID L. TAYLOR

 CENGAGE

Australia • Brazil • Canada • Mexico • Singapore • United Kingdom • United States

Print Reading for Machinists, **6th edition**
David L. Taylor

SVP, GM Skills & Global Product Management:
Jonathan Lau

Product Director: Matthew Seeley

Associate Product Manager: Kelly Lischynsky

Senior Director, Development: Marah Bellegarde

Senior Product Development Manager: Larry Main

Senior Content Developer: Mary Clyne

Product Assistant: Mara Ciacelli

Vice President, Marketing Services:
Jennifer Ann Baker

Marketing Manager: Scott Chrysler

Senior Production Director: Wendy Troeger

Production Director: Andrew Crouth

Senior Content Project Manager: Glenn Castle

Art Director: Angela Sheehan

Cover Image(s): © Shutterstock.com/Galushko Sergey

Composition and Production Service: MPS Limited

For product information and technology assistance, contact us at
Cengage Customer & Sales Support, 1-800-354-9706.

For permission to use material from this text or product, submit all requests online at **www.cengage.com/permissions.**
Further permissions questions can be e-mailed to
permissionrequest@cengage.com.

Library of Congress Control Number: 2017953752

ISBN: 978-1-285-41961-9

Cengage
200 Pier 4 Boulevard
Boston, MA 02210
USA

Cengage is a leading provider of customized learning solutions with employees residing in nearly 40 different countries and sales in more than 125 countries around the world. Find your local representative at **www.cengage.com.**

To learn more about Cengage platforms and services, visit **www.cengage.com.**

To register or access your online learning solution or purchase materials for your course, visit **www.cengage.com.**

Notice to the Reader
Publisher does not warrant or guarantee any of the products described herein or perform any independent analysis in connection with any of the product information contained herein. Publisher does not assume, and expressly disclaims, any obligation to obtain and include information other than that provided to it by the manufacturer. The reader is expressly warned to consider and adopt all safety precautions that might be indicated by the activities described herein and to avoid all potential hazards. By following the instructions contained herein, the reader willingly assumes all risks in connection with such instructions. The publisher makes no representations or warranties of any kind, including but not limited to, the warranties of fitness for particular purpose or merchantability, nor are any such representations implied with respect to the material set forth herein, and the publisher takes no responsibility with respect to such material. The publisher shall not be liable for any special, consequential, or exemplary damages resulting, in whole or part, from the readers' use of, or reliance upon, this material.

Printed in the United States of America
Print Number: 07   Print Year: 2021

CONTENTS

iv   Contents

UNIT 28 Assembly Drawings

UNIT 29 Welding Symbols

UNIT 30 Geometric Tolerances: Datums

UNIT 31 Geometric Tolerances: Location and Form

PREFACE

Most technical professions today require employees to read and interpret industrial prints. Metalworking, quality control, product engineering, process planning for numerical control, computer programming for computer-aided drafting and manufacturing systems, and inspection are just some of the careers that use technical drawings extensively. Students preparing for such careers must strive for excellence in reading and interpreting such drawings quickly and accurately. *Print Reading for Machinists*, 6th edition provides all the basic information a beginning student needs to become skilled at print interpretation.

Features of the New Edition

This resource is designed to present a logical progression of print reading principles in short units of instruction followed by immediate practical application. Each unit contains lessons, examples, review questions, and practice drawings that support the skill development students need most to succeed in the machine trades field.

The basic principles for representing information on a drawing are presented in 31 units. Each unit provides a thorough explanation of specific principles in an easy-to-read style. More than 250 line drawings are provided to illustrate and apply each principle.

To ensure that the student understands industrial practices, 27 end-of-unit assignment drawings are included. The information contained in each unit will enable the student to complete the assignment drawing and answer a series of questions. Additional references are not required to complete assignments.

For ease of learning, these drawings start with relatively simple designs and progress in complexity. As students master new principles and perfect their interpretive skills, the drawings keep pace by providing increasingly challenging assignments.

In addition to assignments relating to the reading of prints, 10 sketching assignments are included to help develop the ability to provide a quick and accurate freehand drawing of a part to be manufactured. *Print Reading for Machinists* conforms to the latest standard of the American National Standards Institute (ANSI), including ASME Y14.5M-2009. The information contained in Unit 29, "Welding Symbols," conforms to the standards of the American Welding Society. The appendices include a review of basic math principles applied to print reading, descriptions of the use of precision measuring tools, a selected list of ANSI abbreviations used on industrial drawings, and assorted handbook tables for quick reference.

MindTap for Blueprint Reading

MindTap® Blueprint Reading for Taylor's *Print Reading for Machinists*, 6th edition is a new digital learning solution that can power students from memorization to mastery. It gives instructors complete control of their courses—to provide engaging content, to challenge every individual, and to build confidence. Customize the interactive assignments and assessments, emphasize the most important topics, or add your own material and notes in the eBook.

Instructor Resources

A robust suite of Instructor Resources is available at the Instructor Companion Website, including an Instructor's Guide with answers to each assignment in the text, PowerPoint lecture slides, and Cengage Testing Powered by Cognero®.

About the Author

David L. Taylor is a former Journeyman Tool and Die Maker with more than 20 years' experience in vocational-technical training. He holds a Master of Science degree in Adult Education from Penn State University and a Bachelor of Science degree in Vocational-Technical Education from the State University of New York at Buffalo. Mr. Taylor has taught courses in machine trades, print reading, and design at Erie County BOCES, Lewis County BOCES, Jamestown Community College, and Ivy Tech State College. Mr. Taylor is the author of four blueprint reading texts published by Cengage.

Acknowledgments

The author and publisher would like to acknowledge the following reviewers for their contributions to this edition:

John Battista, Prosser Career Academy, Chicago, Illinois
Andrew Klein, Reading Muhlenberg Career and Technology Center, Reading, Pennsylvania

The publisher especially acknowledges **Larry Lichter** for his patient and detailed technical edit of the work during the drafting stages.

UNIT 1
INDUSTRIAL DRAWINGS

INTRODUCTION

One of the oldest forms of communication between people is the use of a drawing. A **_drawing_** is a means of providing information about the size, shape, or location of an object. It is a graphic representation that is used to transfer this information from one person to another.

Drawings play a major role in modern industry. They are used as a highly specialized language among engineers, designers, and others in the technical field. These industrial drawings are known by many names. They are called mechanical drawings, engineering drawings, technical drawings, or working drawings. Whatever the term, their intent remains the same. They provide enough detailed information so that the object may be constructed.

Engineers, designers, and drafting technicians commonly produce drawings using computer-aided design and drafting equipment (CAD). The application of computer technology has led to greater efficiency in drawing production and duplication. CAD systems have rapidly replaced the use of mechanical tools to produce original drawings.

COMPUTER-AIDED DESIGN AND DRAFTING

Computer-aided design or **_computer-aided drafting (CAD)_** systems are capable of automating many repetitive, time-consuming drawing tasks. The present technology enables the drafter to produce or reproduce drawings to any given size or view. Three-dimensional qualities may also be given to a part, thus reducing the confusion about the true size and shape of an object. Figure 1.1 shows a typical drawing produced with the help of a computer-aided design system.

CAD systems usually consist of three basic components: (1) hardware, (2) software, and (3) operators or users. The hardware includes a processor, a display system, keyboard, plotter, and digitizer (often called a "mouse"). Software includes the programs required to perform the design or drafting function. Software packages are available in many forms, depending upon the requirements of the user.

The CAD processor is actually the computer or "brains" of the system. The keyboard, which looks very much like a typewriter, is used to place commands into the processor. The commands or input are then displayed graphically on the system display screen. This screen is commonly a liquid crystal display (LCD) or flat cathode ray tube (CRT). The digitizer or mouse is used to create graphic images for display on the screen. The plotter is a printer that produces hard copies of a design in print form.

Industrial drawings are usually produced on a paper material called **_Vellum_** or on a polyester film material known as **_Mylar_**. Mylar is a clear polyester sheet that has a matte finish on one or both sides. The matting provides a dull, granular drawing surface well-suited for pencil or ink lines. Mylar is preferred over vellum in some applications because it resists bending, cracking, and tearing. A completed industrial drawing is known as an original or master drawing.

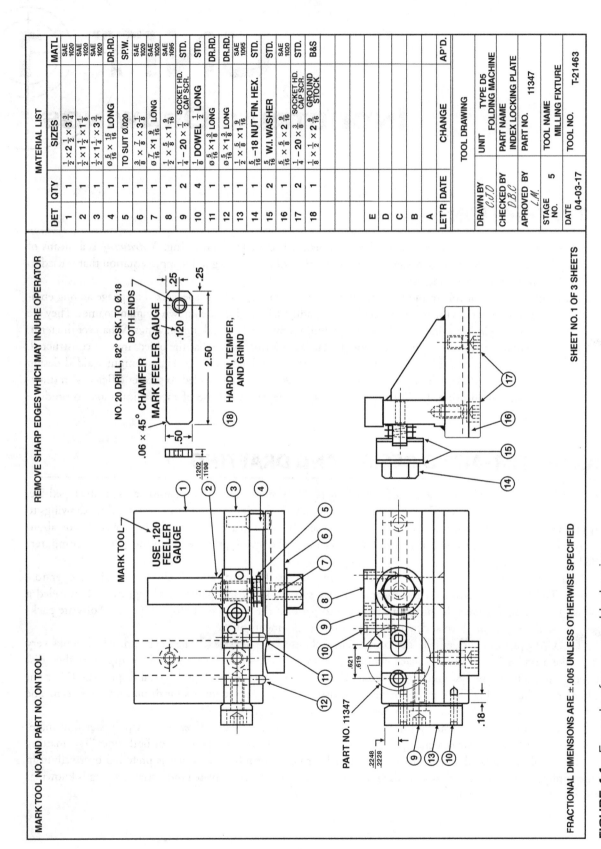

FIGURE 1.1 Example of an assembly drawing.

BLUEPRINTS

Because some original drawings are delicate, they seldom leave the drafting room. They are carefully handled and filed in a master file of originals. When a copy of an original is required, a print is made. The term used for the process of reproducing an original is known as ***blueprinting***. The earliest form of blueprinting produced white line, blue background reproductions. This early process, which was developed in England more than 100 years ago, has since changed. Modern reproductions produce a dark line, white background duplication simply called a ***print***. However, the term ***blueprint*** is still widely used in industry and has been included in the title of this text.

INTERPRETING INDUSTRIAL DRAWINGS

Industrial drawings and prints are made for the purpose of communication. They are a form of nonverbal communication between a designer and builder of a product. Industrial drawings are referred to as a universal language. It is a language that can be interpreted and understood regardless of country. Also, drawings and prints become part of a contract between parties buying and selling manufactured components.

A picture or photograph of an object would show how the object appears. However, it would not show the exact size, shape, and location of the various parts of the object.

Industrial drawings describe size and shape and give other information needed to construct the object. This information is presented in the form of special lines, views, dimensions, notes, and symbols. The interpretation of these elements is called ***print reading***.

PRINT REPRODUCTION PROCESSES

There are several methods available for reproducing drawings.

Chemical Process

The ***ammonia process***, often called the ***diazo process***, is a common method of print reproduction. To produce a copy, the original is placed on top of a light-sensitive print paper. Both the original and the print are fed into the diazo machine and exposed to a strong ultraviolet light. As the light passes through the thin original, it burns off all sensitized areas not shadowed by lines. The print paper is then exposed to an ammonia atmosphere. The ammonia develops all sensitized areas left on the print paper. The result is a dark line reproduction on a light background.

Silver Process

The ***silver process*** is actually a photographic method of reproduction. This process is often referred to as microfilming or photocopying. This method is rapidly gaining popularity in industry due to storage and security reasons. The most common procedure followed is to photograph an original drawing to gain a microfilm negative. The negative is then placed on an aperture card and labeled with a print number. Duplicates of the microfilm are produced with the aid of a microfilm printer using sensitized photographic materials. Enlarged or reduced prints can be produced using this process. The aperture cards containing the microfilm are very small. Therefore, cataloging and filing take very little room for storage. They are also much easier to handle than the delicate originals, which must be kept in large files. Microfilming is often done for security reasons. As many as 200 prints may be placed on one roll of microfilm. They may then be placed in a vault or other secure area.

Electrostatic Process

The *electrostatic process* has gained in popularity for industrial drawing reproduction. Although once limited to reproducing documents and small drawings, new machines have been developed that allow large drawing duplication. The electrostatic process, commonly known as xerography, uses a zinc-coated paper that is given an electrostatic charge. The zinc coating is sensitive to ultraviolet light when exposed. Areas shadowed by lines on the original produce a dark line copy.

CAD Process

As previously described, the **CAD process** uses computer technology to automate many drawing tasks, and to file and store original drawings electronically. One advantage of a CAD system is its ability to rapidly access stored drawings for reproduction or when a revision is required by simply sending a message from the CAD processor to an output device called a printer or printer/plotter, Figure 1.2.

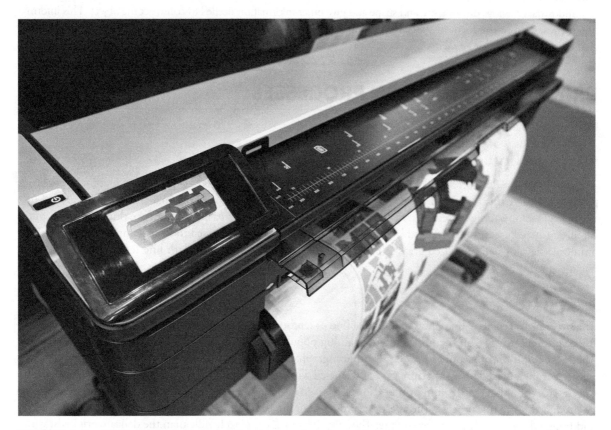

FIGURE 1.2   CAD printer/plotter. iStock.com/sergeyryzhov

ASSIGNMENT: UNIT 1 REVIEW QUESTIONS

1. List two other names commonly given to industrial drawings.

   a. _____

   b. _____

2. Industrial drawings should provide enough information so that the object can be

   _____.

3. The paper material on which original drawings are produced is called _____.

4. A completed industrial drawing is known as a master drawing or _____.

5. Master drawings
   a. Are provided to the machine builder.

   b. Seldom leave the drafting room.

   c. Are developed by the master drafter.

   d. Are always drawn on Vellum.

6. What is the term used for reproducing an industrial drawing?

   _____

   _____

7. Industrial drawings are often referred to as _____ language.

8. Industrial drawings are a form of communication that is:
   a. verbal.

   b. nonverbal.

9. Why is a photograph not used to describe an object?

   _____

   _____

10. The light the print paper is exposed to in the diazo process is:
    a. sunlight.

    b. infrared light.

    c. fluorescent light.

    d. ultraviolet light.

11. The silver process is:
    a. seldom used.

    b. a photographic process.

    c. an ammonia process.

    d. a heat process.

12. List two advantages of microfilming.

    a. _____

    b. _____

13. Aperture cards
    a. are small.

    b. contain print information.

    c. are used for filing.

    d. all of the above.

    e. none of the above.

14. The heat process uses a chemically coated paper that is sensitive to:
    a. infrared light.

    b. heat.

    c. ammonia.

    d. ultraviolet light.

15. The electrostatic process uses paper that is sensitive to:
    a. chemicals.

    b. ammonia.

    c. heat.

    d. light.

16. The electrostatic process uses a paper coated with:
    a. carbon.

    b. lead.

    c. iron.

    d. zinc.

17. List three components of a CAD system.
    a. _____

    b. _____

    c. _____

18. The display screen used with a CAD system is called a _____.

19. What is one advantage a CAD system has over conventional drawing methods?

    _____

    _____

20. What CAD output device is used to produce duplicate copies of original CAD drawings?

    _____

UNIT 2
TITLE BLOCKS

All industrial drawings have certain elements in common. They consist of various lines, views, dimensions, notes, and symbols. Other general information is also supplied so that the object may be completely understood. A skilled print reader must learn to interpret and apply all the information provided on the drawing.

TITLE BLOCKS AND TITLE STRIPS

A *title block* or *title strip* is designed to provide general information about the part, assembly, or the drawing itself.

Title blocks are usually located in the lower right-hand corner of the print, Figure 2.1. Title strips may be used on smaller drawing sheets and extend along the entire lower section of the print, Figure 2.2.

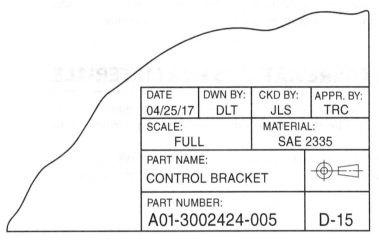

DATE	DWN BY:	CKD BY:	APPR. BY:
04/25/17	DLT	JLS	TRC
SCALE:		MATERIAL:	
FULL		SAE 2335	
PART NAME:			
CONTROL BRACKET			
PART NUMBER:			
A01-3002424-005		D-15	

FIGURE 2.1 Title block.

STANDARD TOLERANCE UNLESS OTHERWISE SPECIFIED		DET.	SHT.	DESCRIPTION	STOCK: FIN. ALLOWED	MAT.	HT. TR.	REQ'D
SPREAD BETWEEN SCREW HOLES MUST BE HELD TO A TOLERANCE OF ±.008 AND SPREAD BETWEEN DOWEL HOLES MUST BE HELD TO A TOLERANCE OF ±.0005		BILL OF MATERIAL ONE						
MILLIMETER	INCH			ABC MACHINE COMPANY				
						JAMESTOWN, NEW YORK		
WHOLE NO. ±0.5	FRACTIONAL ±1/64	TOOL NAME						
1 PLC. DEC ±0.2	2 PLC. DEC ±0.01	FOR:						
2 PLC. DEC ±0.03	3 PLC. DEC ±0.001	OPER:						
3 PLC. DEC ±0.013	4 PLC. DEC ±0.0005	MACHINE:				DATE		
ANGLE ±1/2°		DR.		SCALE	PART No.			
BREAK ALL SHARP CORNERS AND EDGES UNLESS OTHERWISE SPECIFIED		CH.		No. OF SHEETS				
		APP.		SHEET No.	TOOL No.			

FIGURE 2.2 Title strip.

Most companies select a standard form for their drawings that is printed on the original or master drawing. This enables the drafter to simply fill in the required information.

Some of the most common information found in the title block or strip includes the following:

Company name—identifies the company using or purchasing the drawing.

Part name—identifies the part or assembly drawn.

Part number—identifies the number of the part for manufacturing or purchasing information.

Drawing number—is used for reference when filing the original drawing.

Scale—indicates the relationship between the size of the drawing and the actual size of the part. When objects are drawn actual size, the scale would be full scale or 1:1. Large objects are often drawn smaller than actual size. For example, a large part that may not fit on the drawing paper might be drawn half scale or 1:2. Very small objects are often drawn larger than actual size. For instance, the object may be drawn double the actual size of the part. In this example, the scale would be shown as 2:1.

Tolerance—refers to the amount that a dimension may vary from the print. Standard tolerances that apply to the entire print are given in the title block. Tolerances referring to only one surface are indicated near that surface on the print.

Material—indicates the type of material of which the part is to be made.

Heat treat information—provides information as to hardness or other heat treat specifications.

Date—identifies the date the drawing was made.

Drafter—identifies who prepared the original.

Checker—identifies who checked the completed drawing.

Approval—identifies who approved the design of the object.

Change notes or revision—is an area in the block that records for history changes that are made on the drawing. Often revision blocks are located elsewhere on the drawing.

STANDARD ABBREVIATIONS FOR MATERIALS

A variety of materials are used in industry. The drafter or designer must select materials that will best fit the job application. The ability to do this comes from experience and from understanding material characteristics.

To save time and drawing space, material specifications are usually abbreviated on drawings. Table 2.1 describes the most common abbreviations used. Refer to this table as a guide to material abbreviations used later in the text.

Additional tables are found in the Appendix.

TABLE 2.1   Standard Abbreviations for Materials

Alloy Steel	AL STL	Hot-Rolled Steel	HRS
Aluminum	AL	Low-Carbon Steel	LCS
Brass	BRS	Machine Steel	MST
Bronze	BRZ	Malleable Iron	MI
Cast Iron	CI	Nickel Steel	NS
Cold-Drawn Steel	CDS	Stainless Steel	SST
Cold-Finished Steel	CFS	Steel	STL
Cold-Rolled Steel	CRS	Tool Steel	TS
High-Carbon Steel	HCS	Tungsten	TU
High-Speed Steel	HSS	Wrought Iron	WI

PARTS LISTS

A *parts list*, also called a *bill of materials* is often included with the blueprint, Figure 2.3. This list provides information about all parts required for a complete assembly of individual details. The bill of materials is most frequently found on the print that displays the completed assembly and is known as the *assembly drawing*. The assembly drawing is a pictorial representation of a fully assembled unit that has all parts in their working positions.

Additional drawings called *detail drawings* usually accompany the assembly drawing and are numbered for identification. Each assembly detail found in the bill of materials is also provided with a reference number that is used to locate the detail on the detail drawing. Detail drawings give more complete information about the individual units.

Assembly drawings are covered more completely in a later unit of the text.

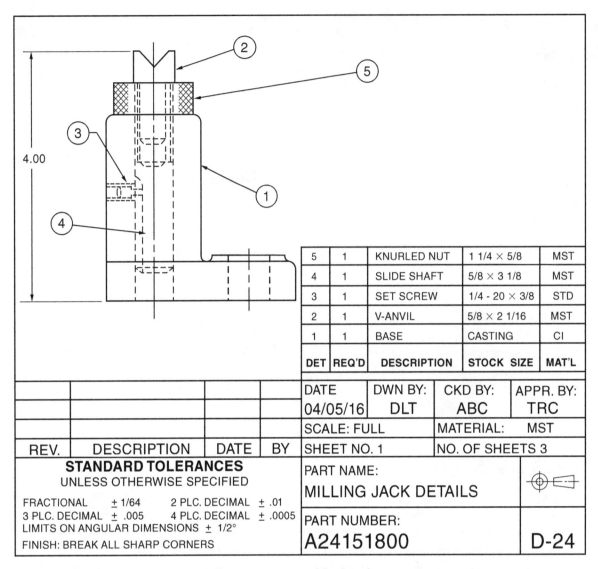

5	1	KNURLED NUT	1 1/4 × 5/8	MST
4	1	SLIDE SHAFT	5/8 × 3 1/8	MST
3	1	SET SCREW	1/4 - 20 × 3/8	STD
2	1	V-ANVIL	5/8 × 2 1/16	MST
1	1	BASE	CASTING	CI
DET	REQ'D	DESCRIPTION	STOCK SIZE	MAT'L

DATE 04/05/16	DWN BY: DLT	CKD BY: ABC	APPR. BY: TRC
SCALE: FULL		MATERIAL: MST	
SHEET NO. 1		NO. OF SHEETS 3	

REV.	DESCRIPTION	DATE	BY

STANDARD TOLERANCES
UNLESS OTHERWISE SPECIFIED

FRACTIONAL   ± 1/64   2 PLC. DECIMAL ± .01
3 PLC. DECIMAL ± .005   4 PLC. DECIMAL ± .0005
LIMITS ON ANGULAR DIMENSIONS ± 1/2°
FINISH: BREAK ALL SHARP CORNERS

PART NAME:
MILLING JACK DETAILS

PART NUMBER:
A24151800   **D-24**

FIGURE 2.3 Example of a parts list on an assembly drawing.

ASSIGNMENT D-1: RADIUS GAUGE

1. What is the name of the part?   _____

2. What material is specified for the Radius Gauge?   _____

3. What is the scale of the drawing?   _____

4. Of what material is the part made?   _____

5. What finish is required?   _____

6. What tolerances are allowed on two-place decimal dimensions?   _____

7. What are the tolerances allowed on three-place decimal dimensions?   _____

8. What are the tolerances allowed on the fractional dimensions?   _____

9. What are the tolerances allowed on the angular dimensions?   _____

10. What is another name for the parts list?   _____

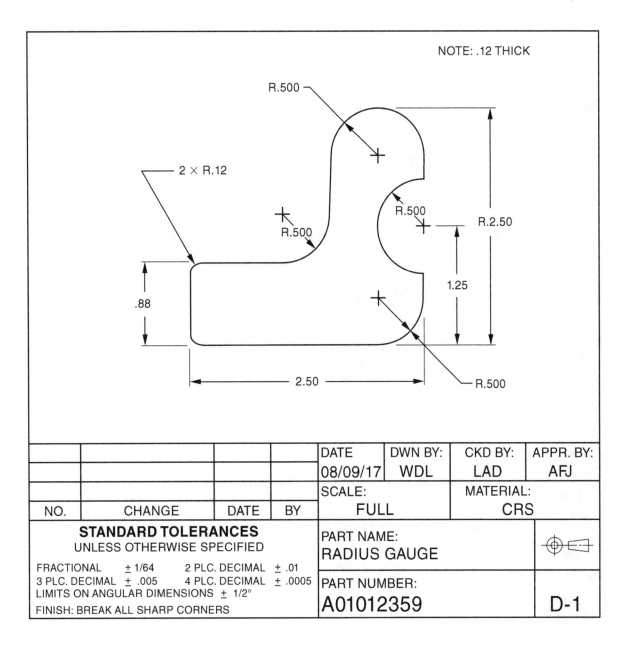

NOTE: .12 THICK

R.500

2 × R.12

R.500

R.500

R.2.50

1.25

.88

2.50

R.500

				DATE	DWN BY:	CKD BY:	APPR. BY:
				08/09/17	WDL	LAD	AFJ
				SCALE:		MATERIAL:	
NO.	CHANGE	DATE	BY	FULL		CRS	

STANDARD TOLERANCES UNLESS OTHERWISE SPECIFIED	PART NAME: RADIUS GAUGE	
FRACTIONAL ± 1/64 2 PLC. DECIMAL ± .01 3 PLC. DECIMAL ± .005 4 PLC. DECIMAL ± .0005 LIMITS ON ANGULAR DIMENSIONS ± 1/2° FINISH: BREAK ALL SHARP CORNERS	PART NUMBER: A01012359	D-1

11. What is the area on the drawing where general information is provided?   _____

12. What is the number used for filing drawings called?   _____

13. Have any changes been indicated on the Radius Gauge?   _____

14. What are copies of originals called?   _____

15. What is the date of this drawing?   _____

UNIT 3

LINES AND SYMBOLS

Various lines on a drawing have different meanings. They may appear solid, broken, thick, or thin. Each is designed to help the blueprint reader make an interpretation. The standards for these lines were developed by the American National Standards Institute (ANSI) and the American Society of Mechanical Engineers (ASME). These lines are commonly known as the Alphabet of Lines, Figure 3.1. Knowledge of these lines helps one visualize the part. Some lines show shape, size, centers of holes, or the inside of a part. Others show dimensions, positions of parts, or simply aid the drafter in placing the various views on the drawing.

This unit describes the most basic lines. The identification of other types of lines will be described in following units.

VISIBLE LINES

Visible lines, also called visible edge or object lines, are heavy, solid lines, Figure 3.2. They generally show the outline, or visible edges of the part.

HIDDEN LINES

Some objects have one or more hidden surfaces that cannot be seen in the given view. These hidden surfaces, or invisible edges, are represented on a drawing by a series of short dashes called *hidden lines*, Figure 3.3.

EXTENSION LINES

Extension lines are thin, solid lines that extend surfaces, Figure 3.4. Extension lines extend away from a surface without touching the object. Dimensions are usually placed between the extension lines.

DIMENSION LINES

Dimension lines are thin, solid lines that show the distance being measured, Figure 3.4. At the end of each dimension line is an *arrowhead*. The points of the arrows touch each extension line. The space being dimensioned extends to the tip of each arrow.

Arrowheads may be open or solid and can vary in size. The size depends mostly on the dimension line weight and blueprint size.

CENTERLINES

Centerlines are thin lines with alternate long and short dashes. They do not form part of the object, but are used to show a location. As the name implies, centerlines indicate centers. They are used to show centers of circles, arcs, or symmetrical parts, Figure 3.5.

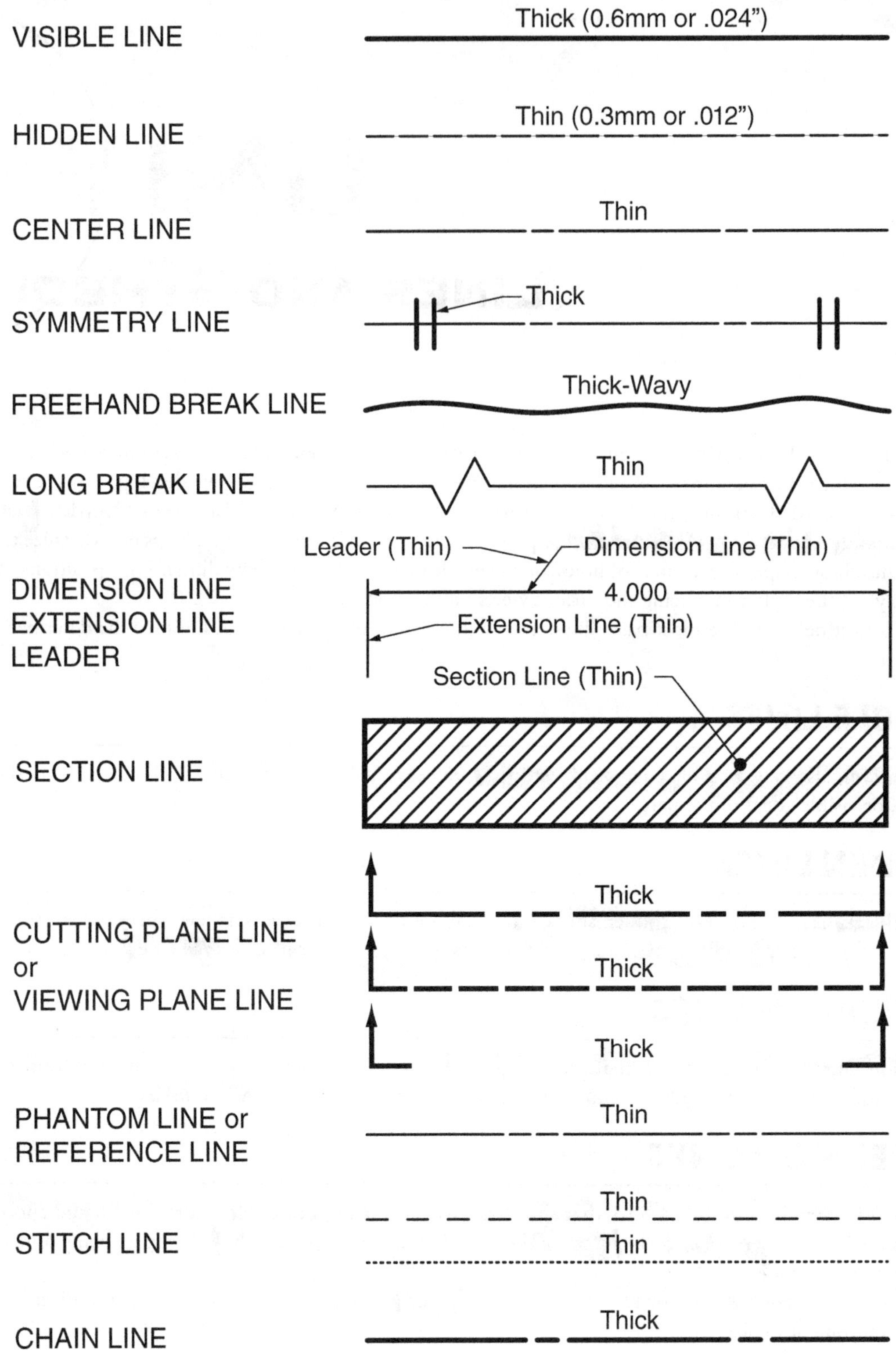

VISIBLE LINE	Thick (0.6mm or .024")
HIDDEN LINE	Thin (0.3mm or .012")
CENTER LINE	Thin
SYMMETRY LINE	Thick
FREEHAND BREAK LINE	Thick-Wavy
LONG BREAK LINE	Thin
DIMENSION LINE EXTENSION LINE LEADER	Leader (Thin) — Dimension Line (Thin) — 4.000 — Extension Line (Thin)
SECTION LINE	Section Line (Thin)
CUTTING PLANE LINE or VIEWING PLANE LINE	Thick / Thick / Thick
PHANTOM LINE or REFERENCE LINE	Thin
STITCH LINE	Thin / Thin
CHAIN LINE	Thick

FIGURE 3.1   Alphabet of lines.

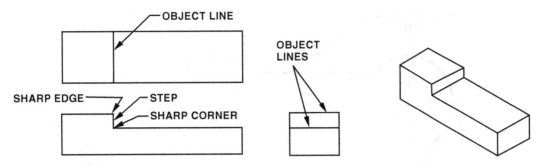

FIGURE 3.2 Visible lines.

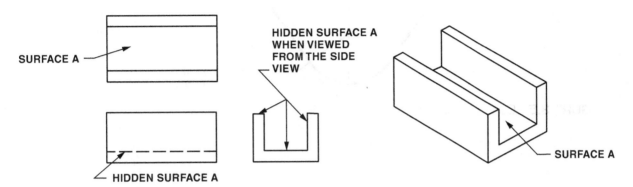

FIGURE 3.3 Hidden surfaces.

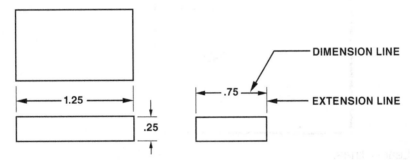

FIGURE 3.4 Extension and dimension lines.

LEADER LINES

Leader lines are similar in appearance to dimension lines. They consist of an inclined line with an arrow at the end where the dimension or surface is being called out. The inclined line is attached to a horizontal leg, at the end of which a dimension or note is provided, Figure 3.6.

SYMMETRY LINES

Symmetry lines are used to show that an object, or a particular feature on an object, is symmetrical. A symmetry line consists of a centerline having two dark, parallel lines, drawn at right angles to each end. This line indicates a plane on the part where each side of the feature is symmetrical, Figure 3.7.

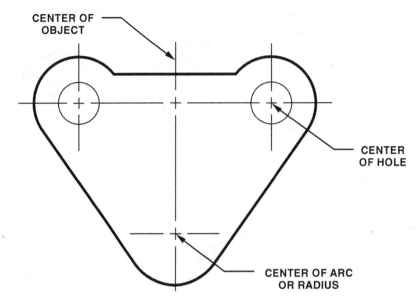

FIGURE 3.5 Centerlines.

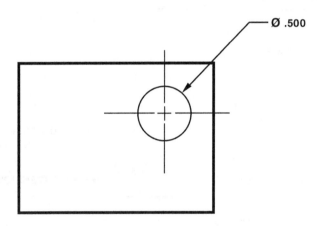

FIGURE 3.6 Leader lines.

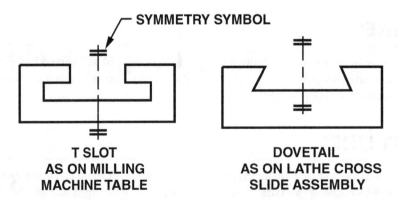

FIGURE 3.7 Symmetry lines.

BREAK LINES

Break lines are often used to show that the view of a part has been shortened or that a portion of the part has been removed because a complete view is not required. Break lines may be dark, wavy, freehand lines to show short breaks (Figure 3.8a), or thin lines with zigzags to show long breaks (Figure 3.8b).

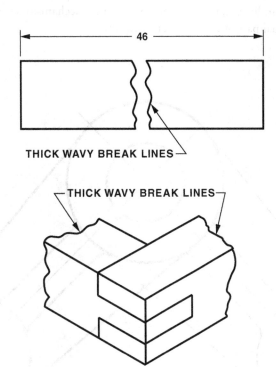

FIGURE 3.8a   Heavy freehand break line for short breaks.

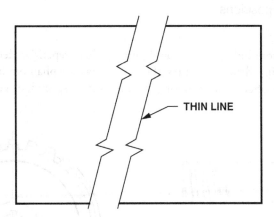

FIGURE 3.8b   Thin zigzag line for long breaks.

PHANTOM LINES

Phantom lines are long dash lines used to indicate the position of an absent part in relation to the view that is shown. (Phantom lines are similar to cutting plane lines but are much lighter, with line breaks that are closer together.) Phantom lines or views drawn in phantom clarify the drawing without the need for additional views.

Phantom views are also used to show mechanisms in alternative positions or how they connect with adjacent parts, Figure 3.9. Phantom lines clarify the operation of the mechanism, even though the view drawn in phantom might not be an actual part of the detail or assembly.

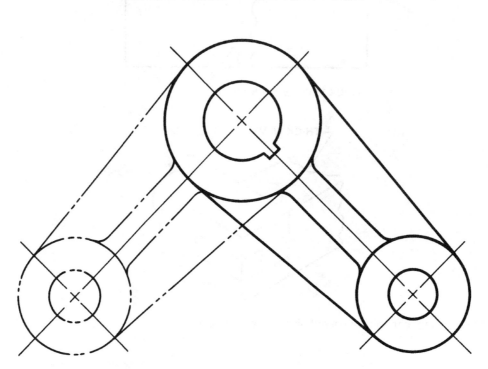

FIGURE 3.9   Alternative positions.

Additional uses of phantom lining include applications where repetitive detail is required on a drawing. Rather than spend time drawing identical features, the drafter may use phantom lines to show the continuation of the feature. This method is often applied to long threaded shafts, springs, or gears, Figure 3.10.

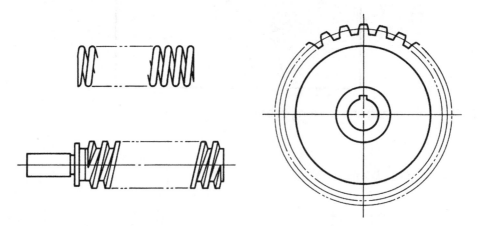

FIGURE 3.10   Use of phantom lines to show repetitive features.

APPLICATION OF SYMBOLS

Revised drawing standards developed by the American National Standards Institute (ANSI) and the American Society of Mechanical Engineers (ASME) are being applied to most modern drawings. These standards encourage the use of symbols to replace words or notes on drawings. This practice reduces drafting time, reduces the amount of written information on the drawing, and helps overcome language barriers. Figure 3.11 shows some common symbols applied to prints. The application of most of these symbols is explained in the appropriate units that follow.

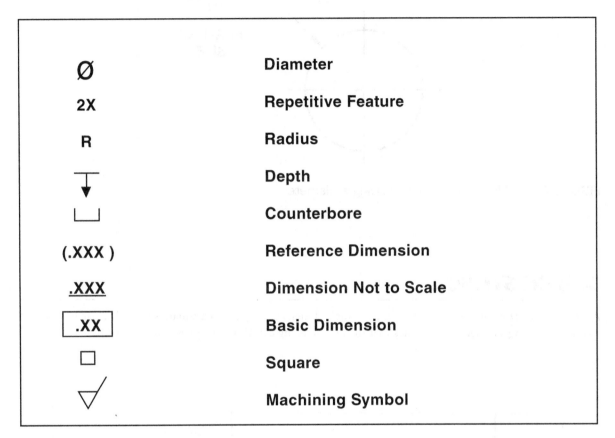

FIGURE 3.11 Standard feature symbols.

DIAMETER SYMBOLS

The former practice was to specify holes or diameters by calling out the hole size, using an abbreviation or letter for the diameter, DIA or D, and a note for the process, Figure 3.12.

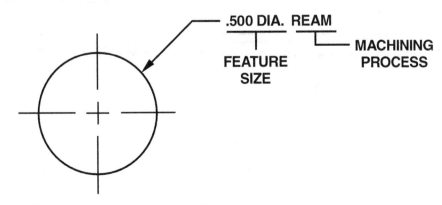

FIGURE 3.12 Old method of specifying a diameter and process.

The new standard for diameter uses the symbol Ø in front of the dimension indicating a diameter and omits the reference to a machining process, Figure 3.13. However, industrial use of the latest standards varies. Many drawings still reflect the older methods.

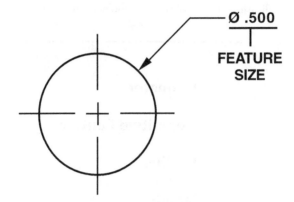

FIGURE 3.13 New method of specifying a diameter.

SQUARE SYMBOL

A *square symbol* is often used to show that a single dimension applies to a square shape. The use of a square symbol preceding a dimension indicates that the feature being called out is square, Figure 3.14.

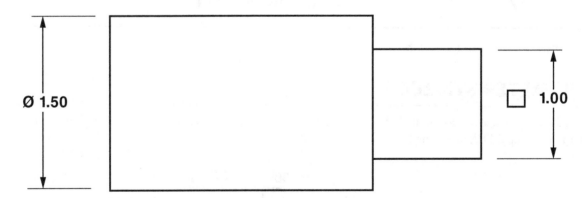

FIGURE 3.14 Application of a square symbol to represent a feature.

SPECIFYING REPETITIVE FEATURES

Repetitive features or dimensions are often specified in more than one place on a drawing. To eliminate the need for dimensioning each individual feature, notes or symbols may be added to show that a process or dimension is repeated.

Holes of equal size may be called out by specifying the number of features required and adding an X. A small space is left between the X and the feature size dimension that follows, Figure 3.15.

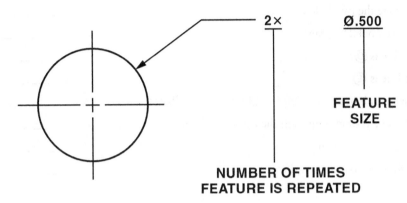

FIGURE 3.15 New method of representing repetitive features.

ASSIGNMENT D-2: TOP PLATE

1. What is the name of the part? _____

2. What is the part number? _____

3. What material is specified for the Top Plate? _____

4. How thick is the part? _____

5. What kind of line is Ⓐ? _____

6. What radius forms the lower end of the part as shown in the front view of the plate? _____

7. How many holes are there? _____

8. What kind of line is Ⓑ? _____

9. How far are the centers of the two holes from the vertical centerline of the piece? _____

10. How far apart are the centers of the two holes? _____

11. What radius is used to form the two large diameters around the ⌀ 1.000 holes? _____

12. What kind of line is Ⓒ? _____

13. What diameter are the two holes? _____

14. What does the symbol 2X mean? _____

15. What kind of line is Ⓓ? _____

16. What kind of line is Ⓔ? _____

17. What is the overall distance from left to right of the Top Plate? _____

18. What kind of a line is drawn through the center of a hole? _____

19. What is the scale of the drawing? _____

20. What special finish is required on the part? _____

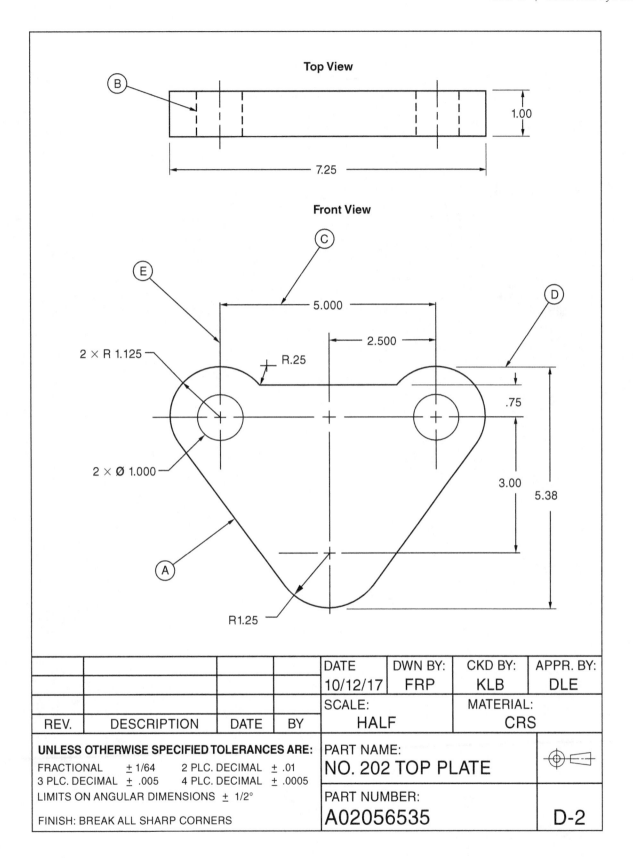

Top View

B

1.00

7.25

Front View

C

E

D

5.000

2.500

2 × R 1.125

R.25

.75

2 × Ø 1.000

3.00

5.38

A

R1.25

				DATE	DWN BY:	CKD BY:	APPR. BY:
				10/12/17	FRP	KLB	DLE
				SCALE:		MATERIAL:	
REV.	DESCRIPTION	DATE	BY	HALF		CRS	

UNLESS OTHERWISE SPECIFIED TOLERANCES ARE:	PART NAME:	
FRACTIONAL ± 1/64 2 PLC. DECIMAL ± .01 3 PLC. DECIMAL ± .005 4 PLC. DECIMAL ± .0005 LIMITS ON ANGULAR DIMENSIONS ± 1/2° FINISH: BREAK ALL SHARP CORNERS	NO. 202 TOP PLATE	⊕ ⊲
	PART NUMBER: A02056535	D-2

UNIT 4

SKETCHING STRAIGHT LINES

Most industrial drawings are made in a drafting room using drafting equipment or computer-aided drafting and design systems. The finished drawings provide the detailed information needed to make the object. Many designs, however, often start with a shop sketch, Figure 4.1.

A shop sketch is a freehand drawing of an object. *Freehand* drawings are made without the aid of drawing instruments, drafting machines, or computers. Shop sketching is often a very important step in the development of an idea.

Shop sketches may be prepared by anyone who needs to communicate an idea in picture form. Engineers, toolmakers, technicians, designers, drafters, and other skilled workers frequently use sketches.

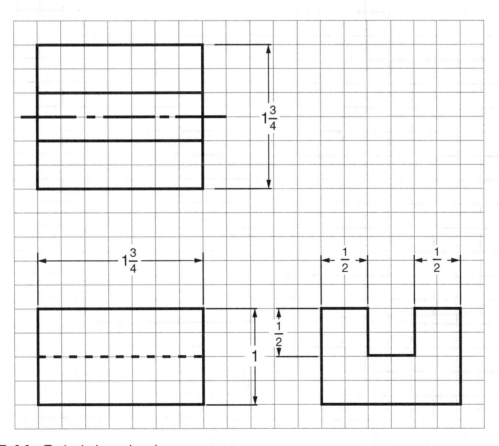

FIGURE 4.1 Typical shop sketch.

EQUIPMENT

Shop sketching requires very little in the way of equipment. The only materials needed are a pencil, an eraser, and paper.

The pencil used should have soft lead such as #2 or HB grade. Soft lead makes a darker-line sketch.

The paper used should be a grid-type, Figure 4.2. Using grid paper helps keep the sketch in the proper proportions and is very helpful for the beginner. An experienced or accomplished sketcher may use unlined paper for sketching.

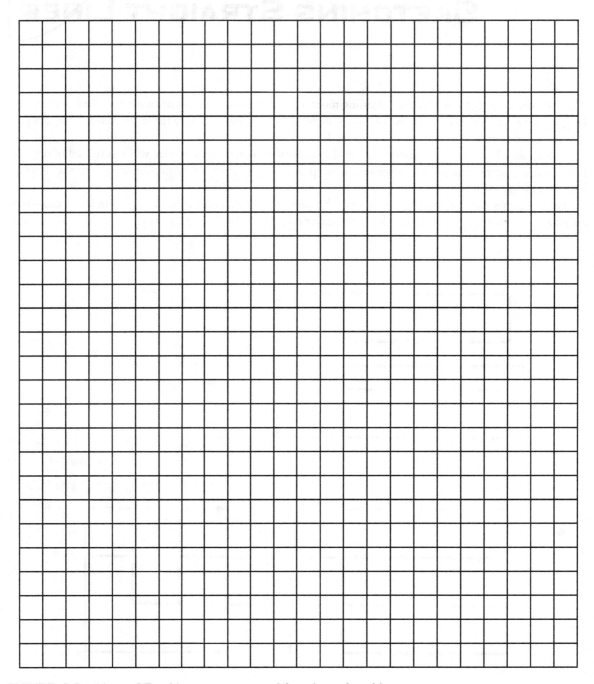

FIGURE 4.2   ¼ or .25 grid-type paper used for shop sketching.

SKETCHING STRAIGHT LINES

The basic straight lines used in sketching are horizontal, vertical, and inclined lines.

Horizontal Lines

Horizontal lines are sketched from left to right between two points. This type of line should be drawn with a movement of the forearm across the paper. To sketch this line:

1. Lay out two points the proper distance apart.

2. Lightly sketch a line between the two points.

3. Darken the lines to the proper weight.

Vertical Lines

Vertical lines are sketched from top to bottom between points. The same procedure should be used as described for horizontal lines. Start at the top point and slowly pull the arm back towards the bottom point.

Inclined Lines

Inclined lines or slanted lines are sketched with the same movement as horizontal and vertical lines. Once the proper angle of incline is determined, a point may be marked off. The line may then be lightly sketched. A suggestion for sketching inclined lines is to turn the paper so that line may be drawn horizontally. Figure 4.3 shows the three basic straight lines.

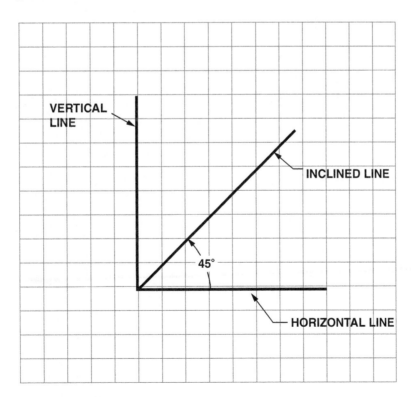

FIGURE 4.3 Basic straight lines.

SKETCH S-1: SHIM PLATE

1. Sketch the shim plate as shown in the following art.

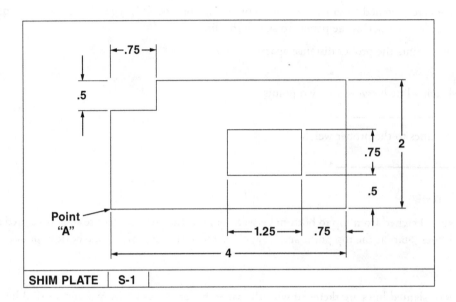

2. On the following grid, start the sketch at Point A about 2 inches from the left-hand margin and 2 inches from the bottom (one square = .25 inch).

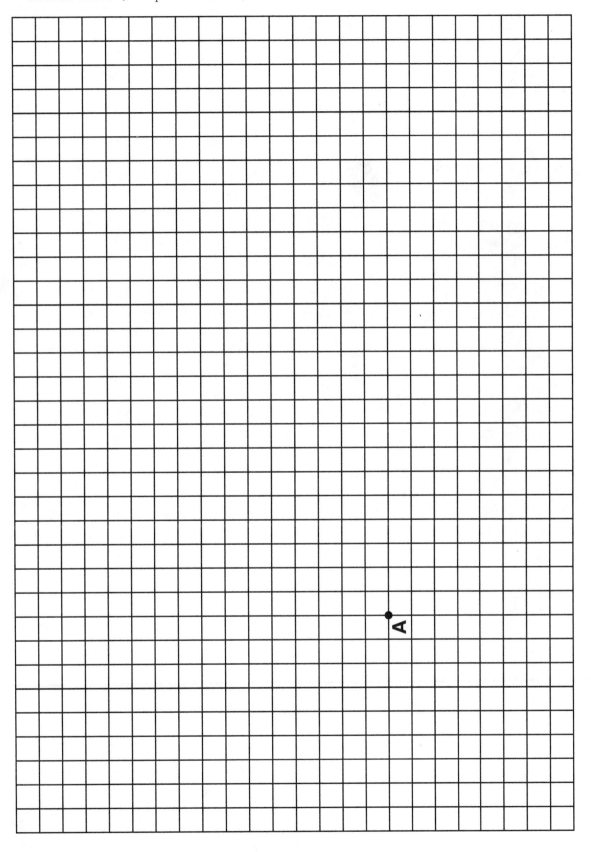

UNIT 5
SKETCHING CURVED LINES

Circles, arcs, and ellipses are the most common curved lines that must be sketched. These types of lines are more difficult to draw accurately and require practice. This unit describes some common methods used to sketch curved lines.

CIRCLES

There is more than one method for sketching circles. Each method uses an aid to help keep the circle round and true size. Perhaps the most common practice for sketching circles is the **square box method**. This involves inscribing the circle in a square that is the proper size for the diameter of the circle. Figure 5.1 shows the proper steps in sketching a circle:

1. Determine the location of the center of the circle.
2. Lightly draw the centerlines of the circle.
3. Mark off the radii of the circle on each centerline.
4. Lightly sketch a square the same diameter as the circle.
5. Starting at the intersection of the centerline and the square, sketch the circle.

ARCS

An **arc** is a part of a circle. The method of sketching an arc is similar to that used for a circle, Figure 5.2:

1. Determine the location of the center of the arc.
2. Lightly draw the centerlines of the arc.
3. Mark off the radii of the arc.
4. Square off the area between the centerlines.
5. Sketch the arc required.

ELLIPSES

An **ellipse** is an oblong-looking circle. The method used to sketch an ellipse is called the **rectangular method**, Figure 5.3:

1. Determine the location of the center of the ellipse.
2. Lightly draw the centerlines of the ellipse.
3. Mark off the major axis of the ellipse on the horizontal centerline.

4. Mark off the minor axis on the vertical centerline.
5. Lightly sketch a rectangle through the points on the centerlines.
6. Sketch the ellipse inside the rectangle.

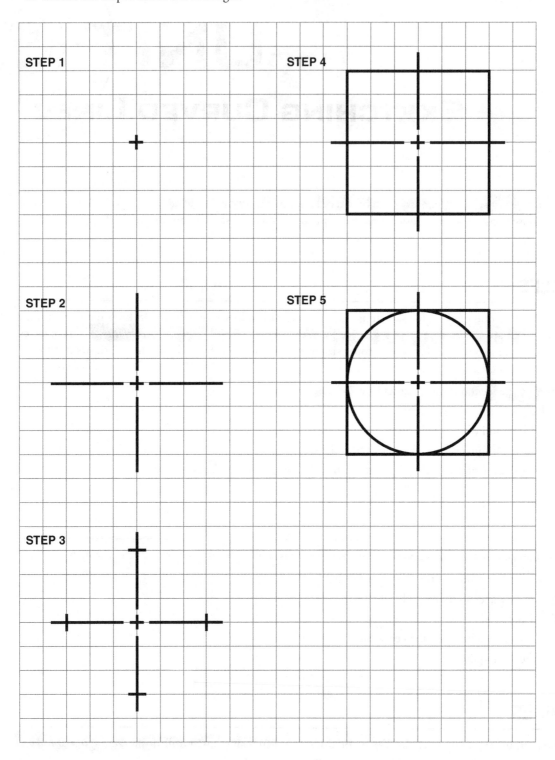

FIGURE 5.1   Sketching a circle.

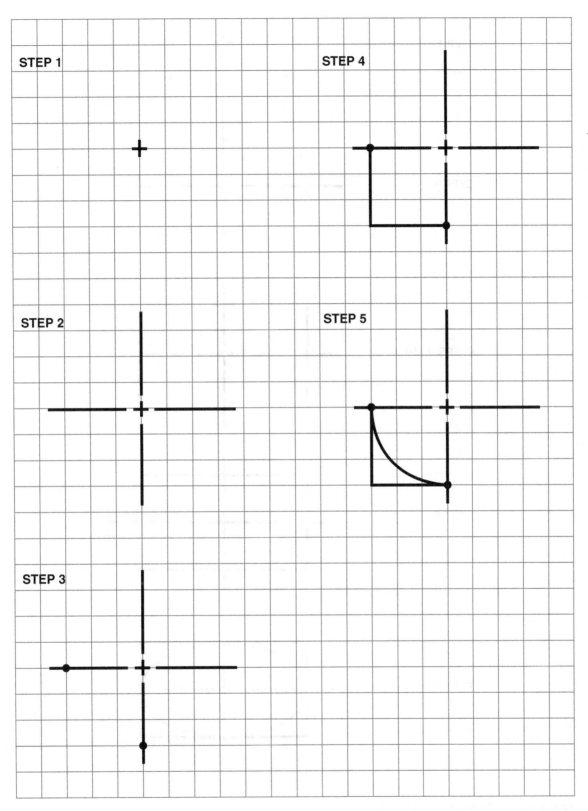

FIGURE 5.2 Sketching an arc.

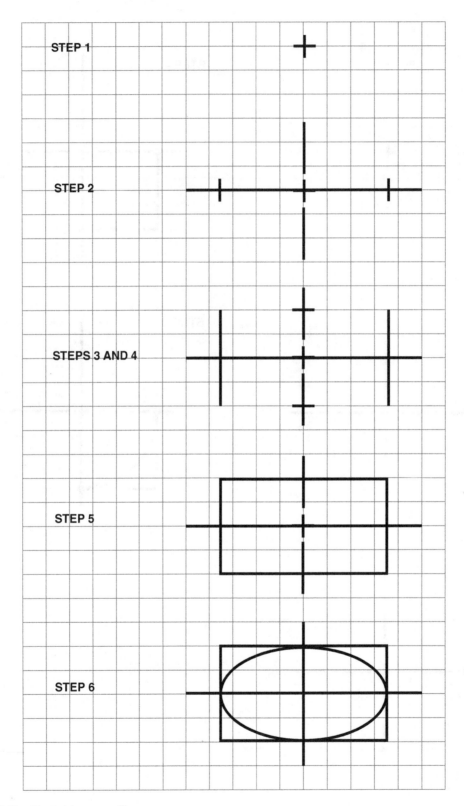

FIGURE 5.3 Sketching an ellipse.

SKETCH S-2: SPACER

Using the centerlines provided in the following grid, sketch the spacer as shown.

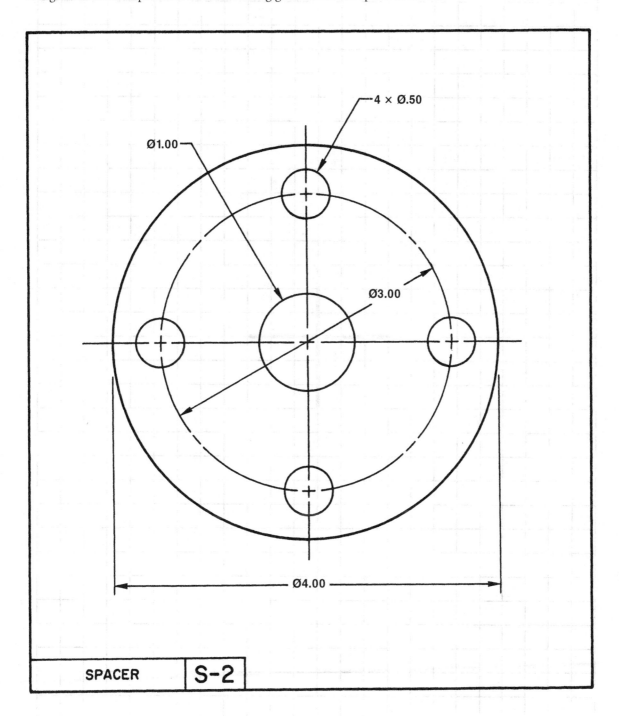

4 × Ø.50

Ø1.00

Ø3.00

Ø4.00

SPACER | S-2

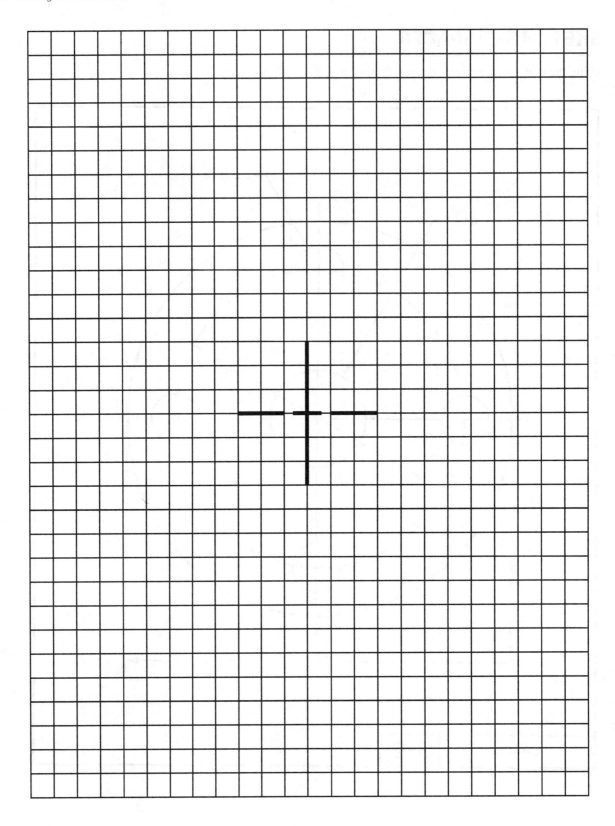

UNIT 6

SKETCHING COMBINATIONS OF LINES

To describe an object with both straight and curved surfaces requires sketching combinations of lines. The principles described in Units 4 and 5 should be applied to combination sketching.

SKETCHING ROUNDED CORNERS

Parts with straight surfaces frequently have rounded corners to remove sharp edges or provide strength. An outside rounded corner is called a *radius* or a *round*. An inside rounded corner is called a ***fillet***. Figure 6.1 describes the steps required to sketch a fillet or round. The procedure is basically the same as described in Unit 5 for sketching arcs. However, the application here is used to connect two straight surfaces.

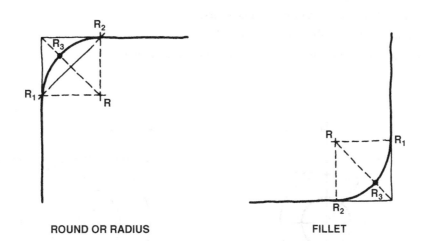

ROUND OR RADIUS   FILLET

FIGURE 6.1   Sketching rounded corners.

CONNECTING CYLINDRICAL SURFACES

Parts such as connecting rods, indexing arms, flanges, and brackets have straight surfaces and cylindrical surfaces. A typical sketch often requires drawing combinations of lines to describe these parts. A good example is the shape description of an open-end adjustable wrench. This type of tool has straight surfaces, curved surfaces, and radii, Figure 6.2.

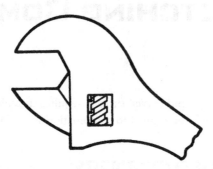

FIGURE 6.2   A open-end adjustable wrench.

Figure 6.3 shows a working drawing for a connecting rod. Figure 6.4 shows the proper steps to connect the cylindrical and flat surfaces:

1. Lay out the centerlines for each cylindrical diameter.
2. Lay out the radii of each diameter on the centerlines.
3. Box in the circular portions of the connecting rod.
4. Sketch the cylindrical ends of the rod and connect each with straight lines.

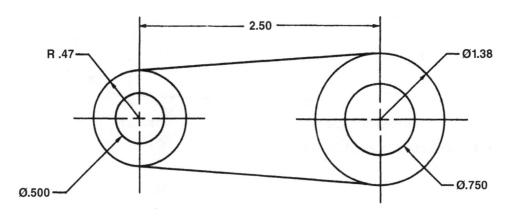

FIGURE 6.3   Connecting rod.

STEP 1

STEP 2

STEP 3

STEP 4

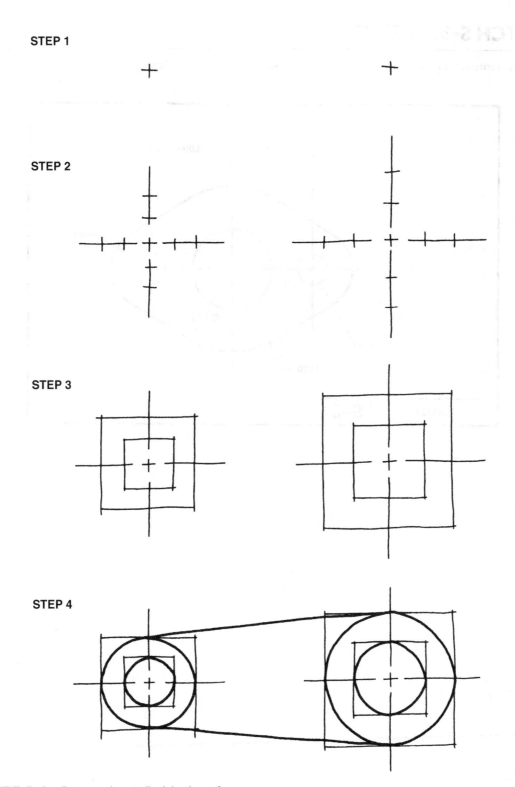

FIGURE 6.4 Connecting cylindrical surfaces.

SKETCH S-3: GASKET

Using the centerlines provided in the following grid, sketch the gasket as shown.

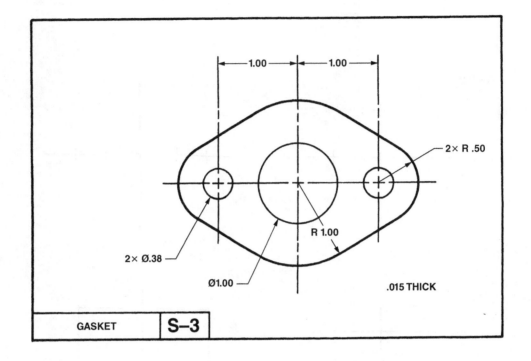

| GASKET | S–3 |

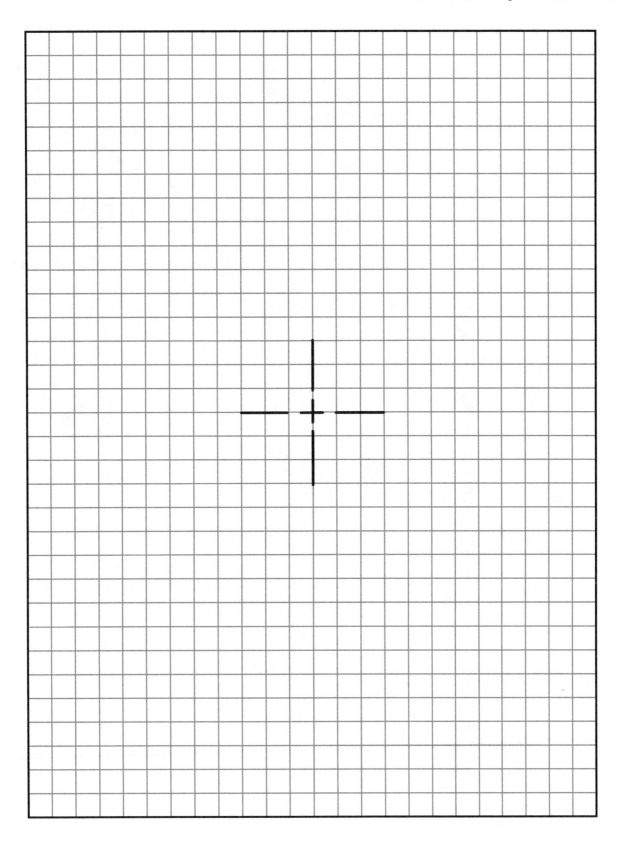

UNIT 7

PICTORIAL SKETCHING

Industrial drawings are representations of three-dimensional objects drawn on a flat plane. The print reader must observe the drawing and interpret the true size, shape, and function of the object. The ability to visualize an object accurately can be simplified by providing the observer with a *pictorial sketch* of the part.

In a pictorial view, called an *axonometric representation* the object is rotated around one or more axes to show combinations of surfaces in one view. This allows individuals with limited technical training to interpret the shape of the object.

ISOMETRIC PROJECTION

Isometric projection is the most frequently used class of pictorial sketch, and it has been used exclusively on the drawings in this text. Objects sketched in isometric provide a descriptive view of more than one surface. In isometric drawings, all three principal angles are of equal size. Additionally, the three principal surfaces are equally inclined to the plane of projection, Figure 7.1.

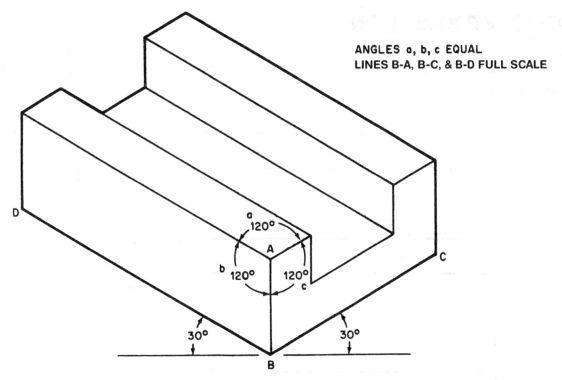

ANGLES a, b, c EQUAL
LINES B-A, B-C, & B-D FULL SCALE

FIGURE 7.1 Isomertic drawing.

True isometric projections must be drawn using a special isometric scale that is two-thirds full size. The angle of inclination in the isometric requires the projection to be drawn at less than true size.

Generally, however, when sketching an isometric, the two-thirds-size rule is disregarded and the principal surfaces are drawn full scale. It is this full-scale representation, together with equal angular axes, that makes isometric projection preferable over other types of axonometric representations.

The true size and shape of a hole cannot be represented accurately on an isometric drawing. The surfaces are inclined to the plane of projection and holes are shown as ellipses, Figure 7.2.

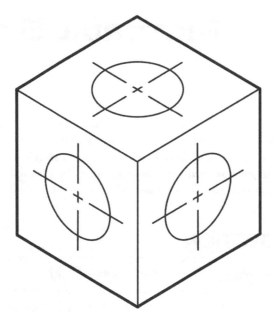

FIGURE 7.2   Holes in isometric.

OBLIQUE PROJECTION

Oblique projection, like axonometric projection, places the object in such a manner that more than one surface is shown in a single view. In an oblique drawing, however, the object is placed with one principal face parallel to the plane of projection, Figure 7.3. The lines that recede from the principal face may be drawn at any convenient angle.

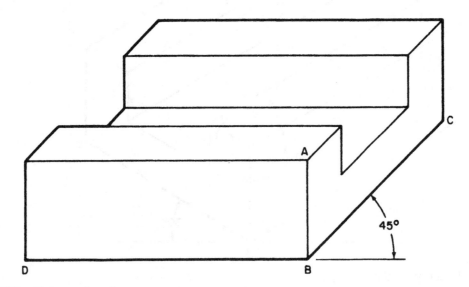

FIGURE 7.3   Oblique drawing.

Generally, a 45-degree angle is used. The angle selected often depends on whether a greater or smaller angle will best show features which appear on receding surfaces.

The lengths of receding lines on oblique drawings are most frequently shown full scale. This, however, leads to some visual distortion of the true size of the object. The degree of distortion may be limited by shortening the length of the receding surfaces to any amount selected, Figure 7.4.

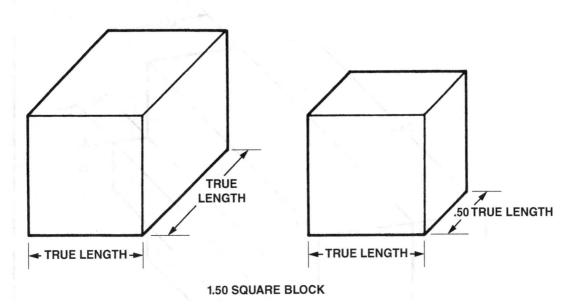

FIGURE 7.4 Oblique drawings.

An advantage of oblique drawing is that all features on a principal surface may be drawn in true size and true shape. Therefore, holes drawn in oblique appear as circles on the principal surface, Figure 7.5.

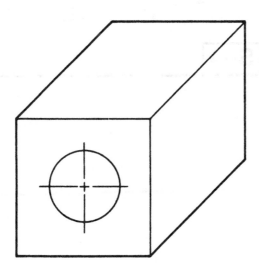

FIGURE 7.5 Holes in oblique.

SKETCH S-4: FORM BLOCK

1A. Sketch an isometric view of the form block using the corner "A" given in the following grid.

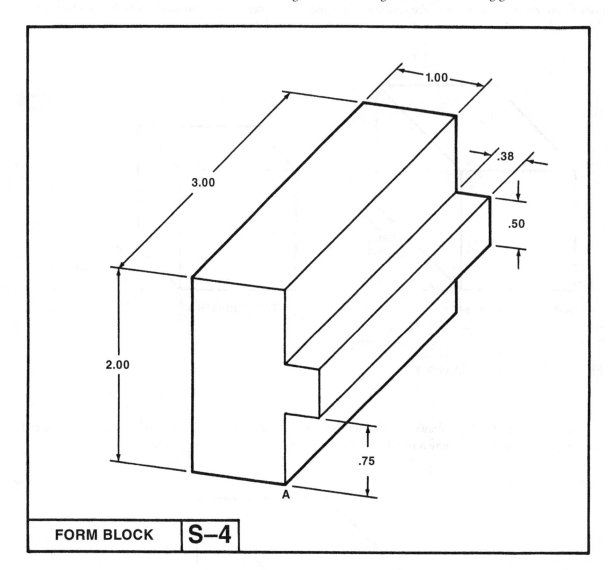

FORM BLOCK   **S–4**

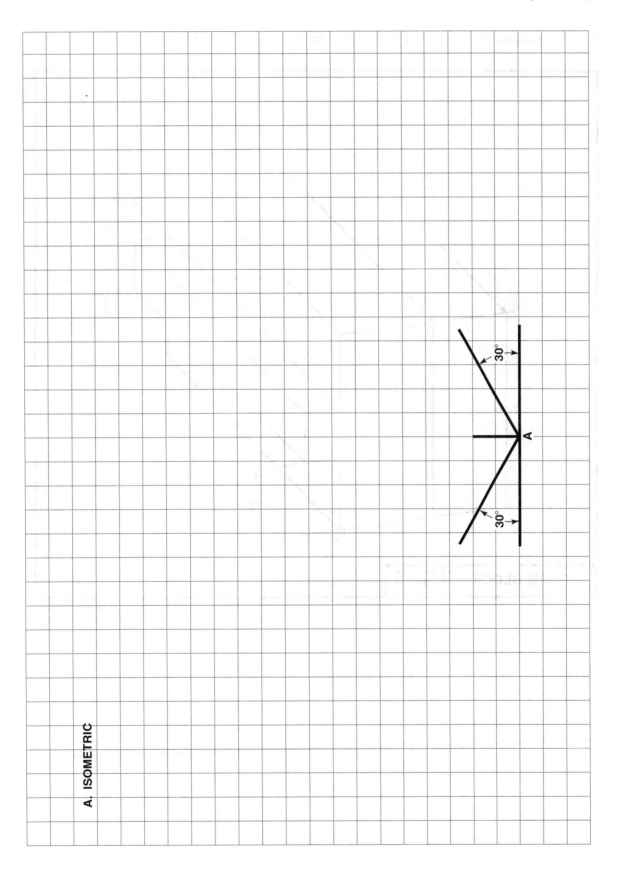

A. ISOMETRIC

1B. Using the corner "A" given in the following grid, sketch an oblique view of the form block.

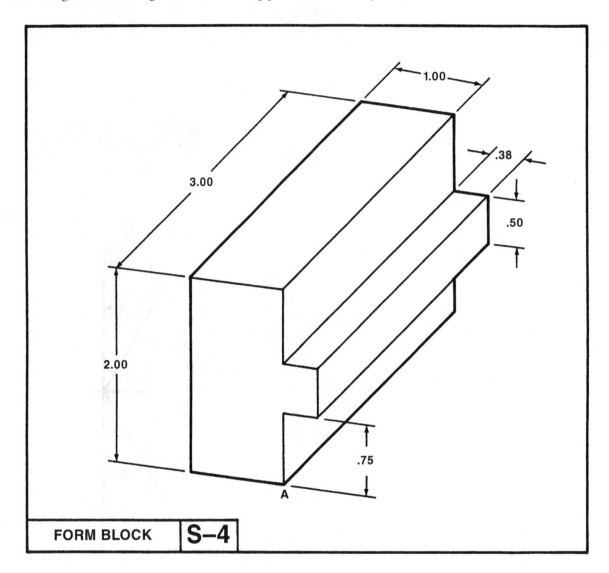

FORM BLOCK   **S–4**

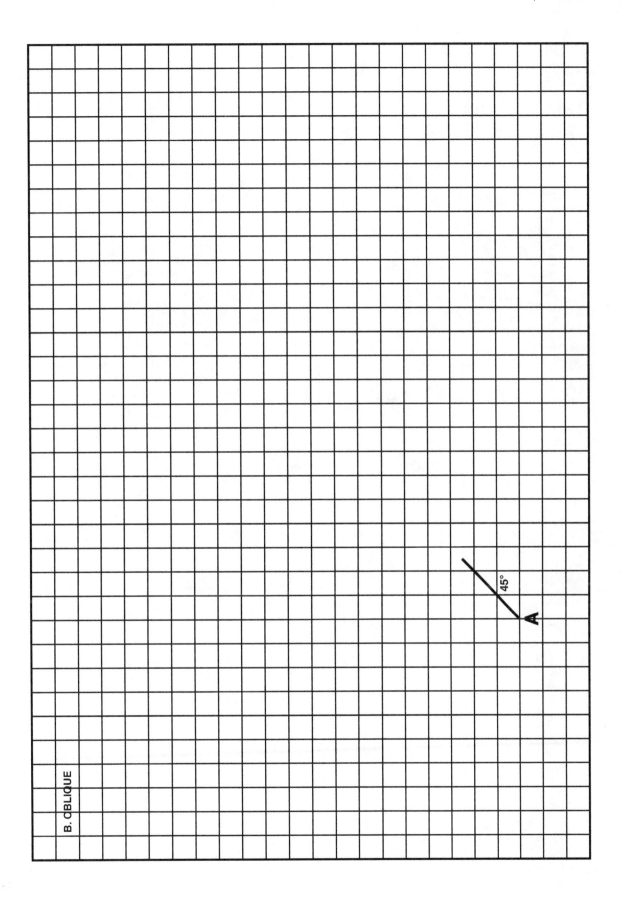

B. OBLIQUE

45°

A

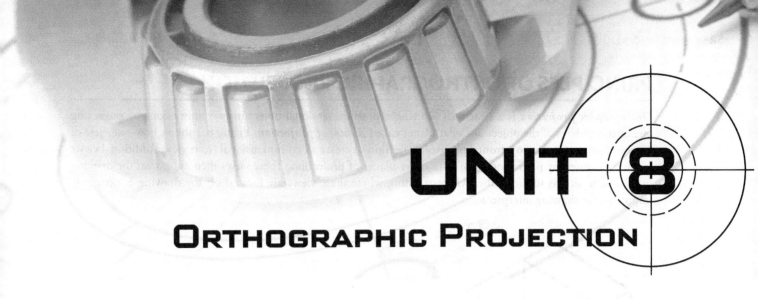

UNIT 8

ORTHOGRAPHIC PROJECTION

WORKING DRAWINGS

Industrial drawings and prints furnish a description of the shape and size of an object. All information necessary for the object's manufacture must be presented in a form that is easily recognized. For this reason, a number of views may be necessary. Each view shows a part of the object as it is seen by looking directly at each one of the surfaces. When all the appropriate notes, symbols, and dimensions are added, it becomes a working drawing. A *working drawing* supplies all the information required to construct the part, Figure 8.1. The ability to interpret a drawing accurately is based on the mastery of two skills. The print reader must:

1. Visualize the completed object by examining the drawing itself.
2. Know and understand certain standardized signs and symbols.

Visualizing is the process of forming a mental picture of an object. It is the secret of successful drawing interpretation. Visualization requires an understanding of the exact relationship of the views to each other. It also requires a working knowledge of how the individual views are obtained through projection. When these views are connected mentally, the object has length, width, and thickness.

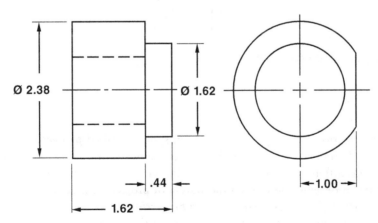

FIGURE 8.1 A working drawing.

PRINCIPLES OF ORTHOGRAPHIC PROJECTION

Orthographic projection is a system of showing a three-dimensional object in two dimensions by projecting one or more views of the object onto flat planes called *planes of projection*. Figure 8.2 shows how one view of an object is projected onto a frontal plane of projection to create a two-dimensional front view. Additional views would be created by projecting them onto other planes of projection. These views then appear on the drawing in relative positions to each other. Understanding where these views are located on the drawing is extremely important for drawing interpretation.

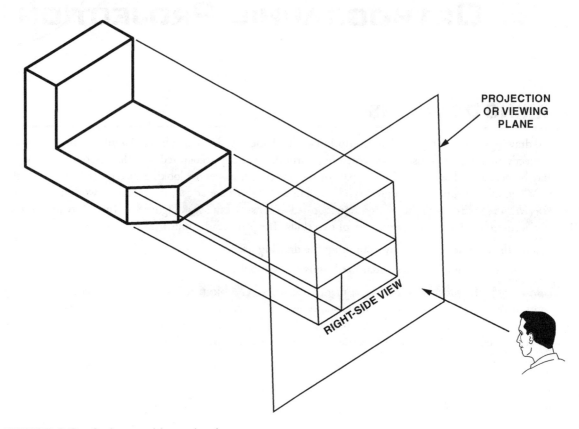

PROJECTION OR VIEWING PLANE

RIGHT-SIDE VIEW

FIGURE 8.2 Orthographic projection.

There are two internationally recognized projection systems (called *projection angles*) used to create orthographic views. Although the views in both systems are nearly the same, the position in which they appear on the drawing is different. Therefore, it is very important for the print reader to know which angle of projection was used to develop the views. One system of projection is called *third angle orthographic projection* and the other is *first angle orthographic projection*. The basic difference between the two is the placement of the projection planes and the placement of the views on the drawing. Imagine a horizontal and a vertical viewing plane intersecting at right angles forming four quadrants. *Third angle projection* places the object in the third quadrant, which is below the horizontal viewing plane and behind the frontal viewing plane. First angle projection places the object in the first quadrant, above the horizontal viewing plane and in front of the frontal plane, Figure 8.3.

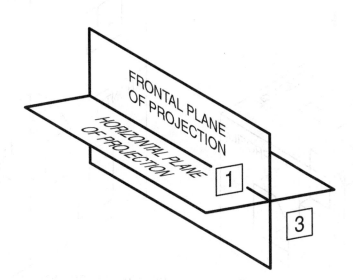

FIGURE 8.3 Projection quadrants.

THIRD ANGLE ORTHOGRAPHIC PROJECTION

Third angle projection is the recognized standard in the United States, Great Britain, and Canada. In this system, the planes of projection are between the observer and the object. To project the principal views of the object, imagine that it is inside a box with transparent sides, as shown in Figure 8.4A. The front view is projected onto the frontal plane in front of the object. The top view is projected onto the horizontal plane directly above the object. The right-side view is projected on the profile plane to the right of the object, as shown in Figure 8.4B. The relative positions of views in third angle orthographic projection are shown in Figure 8.4C.

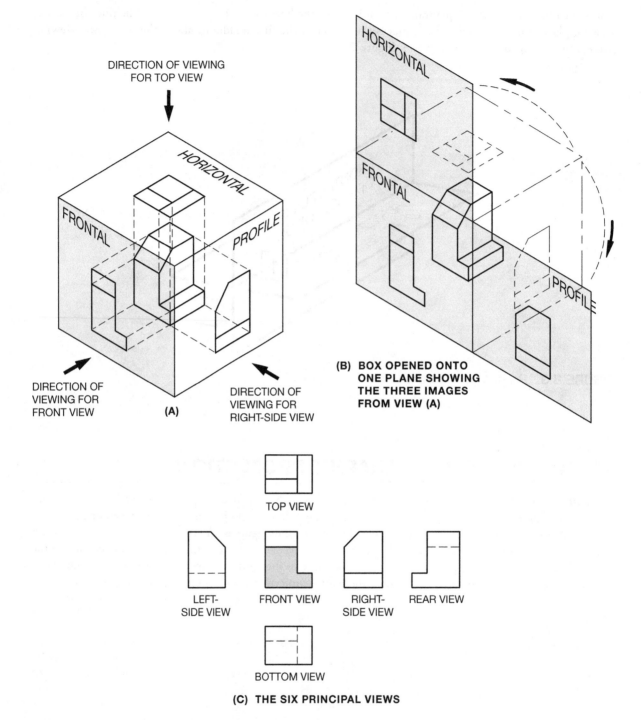

DIRECTION OF VIEWING
FOR TOP VIEW

HORIZONTAL

FRONTAL

PROFILE

DIRECTION OF
VIEWING FOR
FRONT VIEW

(A)

DIRECTION OF
VIEWING FOR
RIGHT-SIDE VIEW

HORIZONTAL

FRONTAL

PROFILE

**(B) BOX OPENED ONTO
ONE PLANE SHOWING
THE THREE IMAGES
FROM VIEW (A)**

TOP VIEW

LEFT-
SIDE VIEW

FRONT VIEW

RIGHT-
SIDE VIEW

REAR VIEW

BOTTOM VIEW

(C) THE SIX PRINCIPAL VIEWS

FIGURE 8.4 Third angle projection.

Note: All drawings in this text are shown using third angle projection.

FIRST ANGLE ORTHOGRAPHIC PROJECTION

First angle projection is used in most European countries as well as Asia. In this system, the object is between the observer and the plane of projection. The views appear as though the observer were looking through the object, as shown in Figure 8.5A. The surface features on the front of the object are projected onto the frontal plane of projection behind the object. The top view is projected onto the horizontal projection plane below the object. The right-side view is projected onto the profile plane on the left side of the object, as shown in Figure 8.5B. The relative positions of views in first angle orthographic projection are shown in Figure 8.5C.

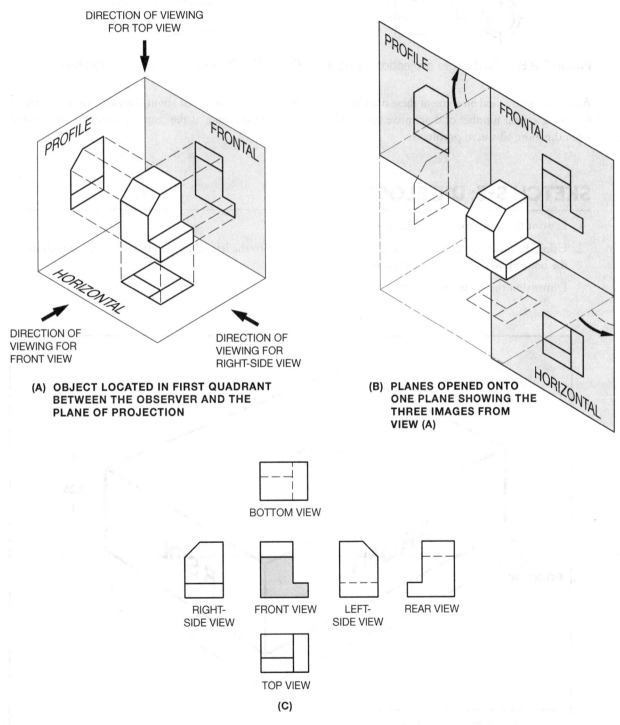

(A) **OBJECT LOCATED IN FIRST QUADRANT BETWEEN THE OBSERVER AND THE PLANE OF PROJECTION**

(B) **PLANES OPENED ONTO ONE PLANE SHOWING THE THREE IMAGES FROM VIEW (A)**

BOTTOM VIEW

RIGHT-SIDE VIEW   FRONT VIEW   LEFT-SIDE VIEW   REAR VIEW

TOP VIEW

(C)

FIGURE 8.5 First angle projection.

ISO PROJECTION SYMBOLS

A **projection symbol** is used on the drawing to show if the views are third angle or first angle projection. This symbol, which was developed by the International Standards Organization (ISO), is found in or near the title block area of the drawing. Figures 8.6 and 8.7 show the standard ISO symbols used to indicate first angle and third angle projection, respectively.

FIGURE 8.6   Third angle projection symbol.   **FIGURE 8.7**   First angle projection symbol.

Note: Letters are used on some of these drawings so that questions may be asked about lines and surfaces without involving a great number of descriptive items. They are learning aids and, as the course progresses, are omitted from the more advanced problems.

SKETCH S-5: DIE BLOCK

1. Lay out the front, top, and right-side views.

2. Using the following grid, start the sketch about ¼ inch from the left-hand margin and about ½ inch from the bottom. Make the views 1 inch apart.

3. Dimension the completed sketch.

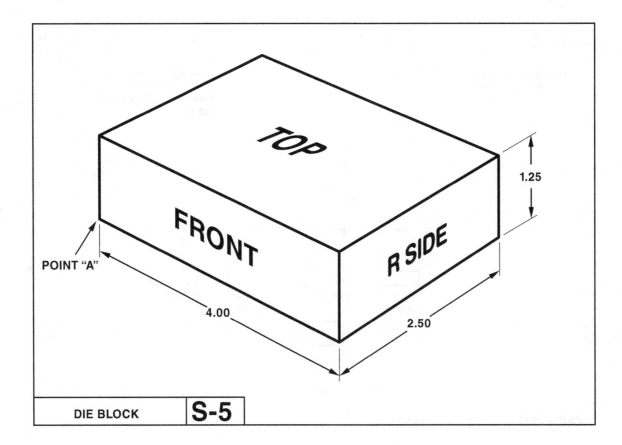

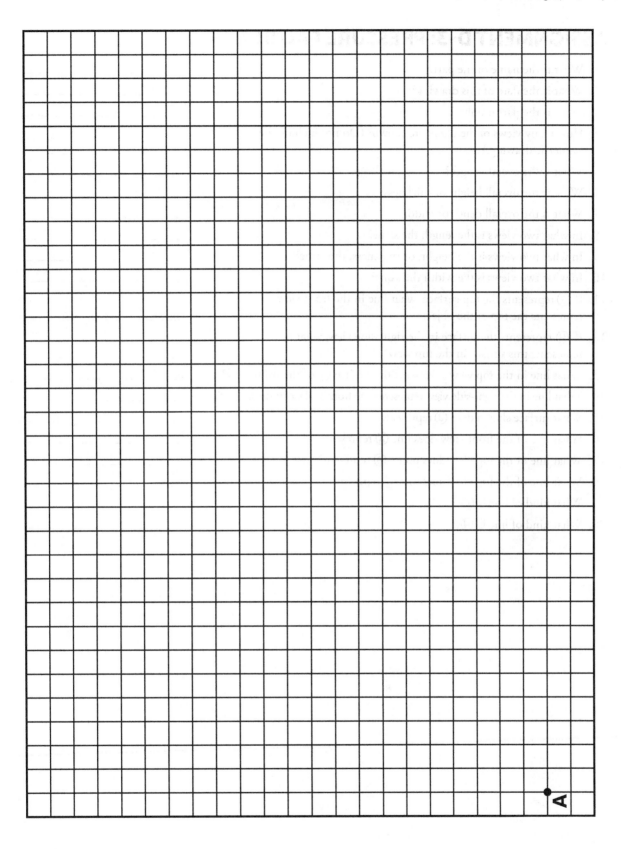

ASSIGNMENT D-3: PRESSURE PAD BLANK

1. What is the name of the part? _____

2. What is the date of this drawing? _____

3. What is the part number? _____

4. How many views of the blank are shown? (Do not include the isometric drawing.) _____

5. What is the overall length? _____

6. What is the overall height or thickness? _____

7. What is the overall depth or width? _____

8. In what two views is the length the same? _____

9. In what two views is the height, or thickness, the same? _____

10. In what two views is the width the same? _____

11. If Ⓕ represents the top surface, what line in the front view represents the top of the object? _____

12. If Ⓗ represents the surface in the right-side view, what line represents this surface in the top view? _____

13. What line in the top view represents the surface Ⓖ of the front view? _____

14. What line in the right-side view represents the front surface of the front view? _____

15. What surface shown does Ⓙ represent? _____

16. What line of the front view does line Ⓐ represent? _____

17. What line of the top view does point Ⓑ represent? _____

18. What line of the top view does line Ⓒ represent? _____

19. What kind of line is Ⓛ? _____

20. What kind of line is Ⓝ? _____

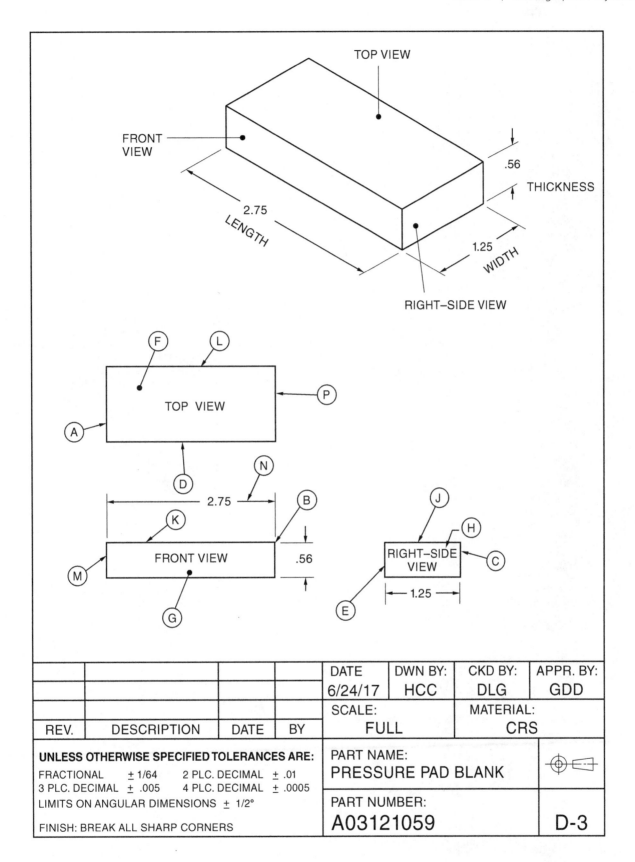

				DATE	DWN BY:	CKD BY:	APPR. BY:
				6/24/17	HCC	DLG	GDD
				SCALE:		MATERIAL:	
REV.	DESCRIPTION	DATE	BY	FULL		CRS	

UNLESS OTHERWISE SPECIFIED TOLERANCES ARE:

FRACTIONAL   ± 1/64   2 PLC. DECIMAL   ± .01
3 PLC. DECIMAL ± .005   4 PLC. DECIMAL ± .0005
LIMITS ON ANGULAR DIMENSIONS ± 1/2°

FINISH: BREAK ALL SHARP CORNERS

PART NAME:
PRESSURE PAD BLANK

PART NUMBER:
A03121059

D-3

UNIT ⊕ 9
THREE-VIEW DRAWINGS

THREE-VIEW DRAWINGS

A common practice when developing an engineering drawing is to display the object using a top, front, and right-side view. These three views typically provide all the shape and size detail needed to produce the part. However, there are some instances where other views may be selected to provide greater clarity.

Figure 9.1 shows how a typical three-view drawing is developed. Figure 9.2 shows how the object would appear on the final drawing. The projection lines shown on both figures are provided to show how the surfaces

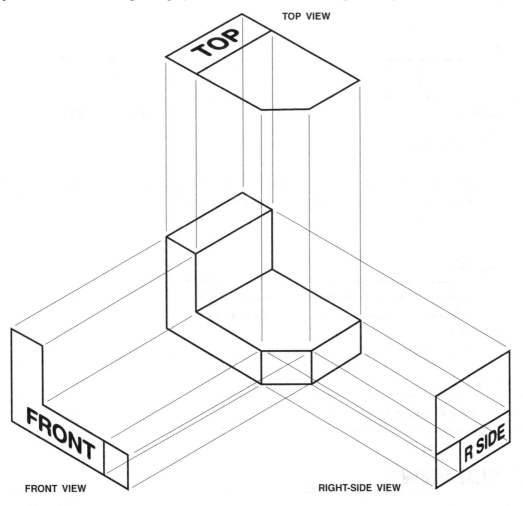

FIGURE 9.1 Projecting the three principal views.

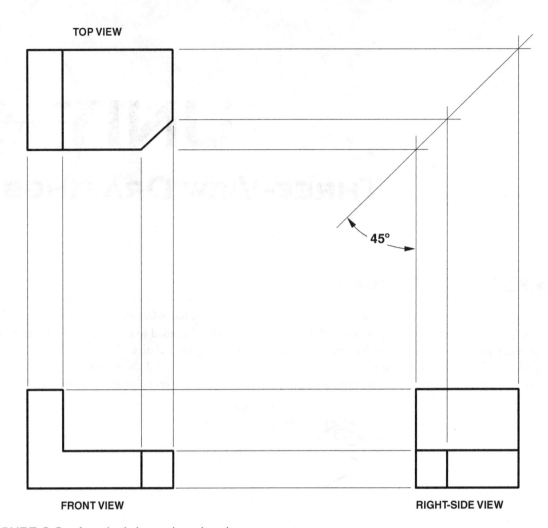

FIGURE 9.2 A typical three-view drawing.

of the object are projected and the alignment of those surfaces between the views. The miter line in Figure 9.2 demonstrates view alignment and equal spacing between the views. Projection lines do not appear on a finished drawing and are used here for illustration only.

FRONT VIEW

The *front view* of an object is the view that shows the greatest detail of the part. It generally is the view of the object that will provide the most descriptive shape information. It is not necessarily the front of the part as it is used in actual operation.

TOP VIEW

The *top view* is located directly over and in line with the front view. It is drawn as if the drafter were standing over the object and looking straight down.

RIGHT-SIDE VIEW

The *right-side view* is the most common side view used in a three-view drawing. It is drawn to the right and in line with the front view.

BOTTOM VIEWS

Bottom views are not ordinarily used in standard drawing practice and thus may be considered a special view. Prints with top, front, and right-side views, or some combination of these views, are most common.

Bottom views are used when it is necessary to bring out details not clearly shown in other views. The bottom view should include only those lines that are necessary to complete the description of the object, Figure 9.3. By the same rule, the lines shown on the bottom view may be omitted from the top view.

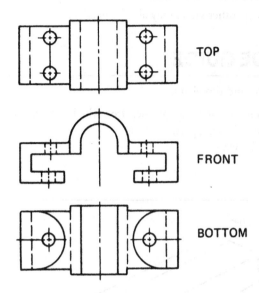

FIGURE 9.3 Application of the bottom view.

PARTIAL VIEWS

A **partial view** is one on which only part of an object is shown. Sufficient information should be given in the partial view to complete the description of the object. Figure 9.4 illustrates an example of proper use of a partial view. A break line is provided to show where the view ends.

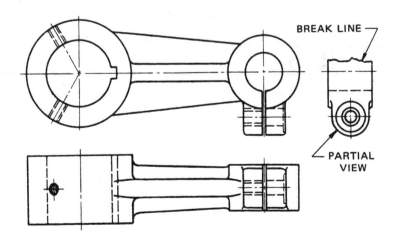

FIGURE 9.4 Partial view.

Partial views provide the following advantages:

- They save drawing time.
- They conserve space that might otherwise be required to draw the object completely.
- They sometimes permit a larger scale view. This allows details to be brought out more clearly.

OTHER VIEWS

If the object is even more complex, other views may also be needed.

SKETCH S-6: SLIDE GUIDE

1. Lay out the front, right-side, and top views.
2. Using the following grid, start the drawing .50 inch from the left-hand margin and about .25 inch from the bottom. Make the views .75 inch apart.
3. Dimension the completed drawing.

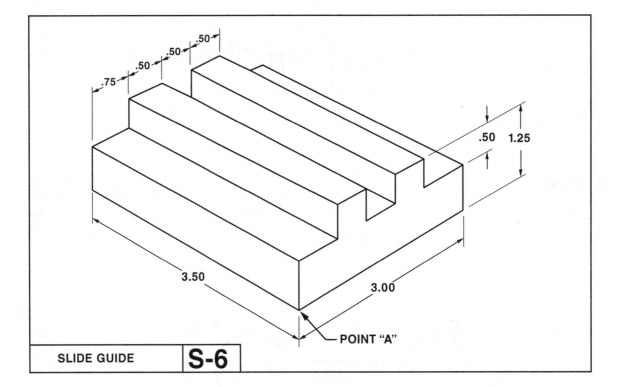

SLIDE GUIDE   **S-6**

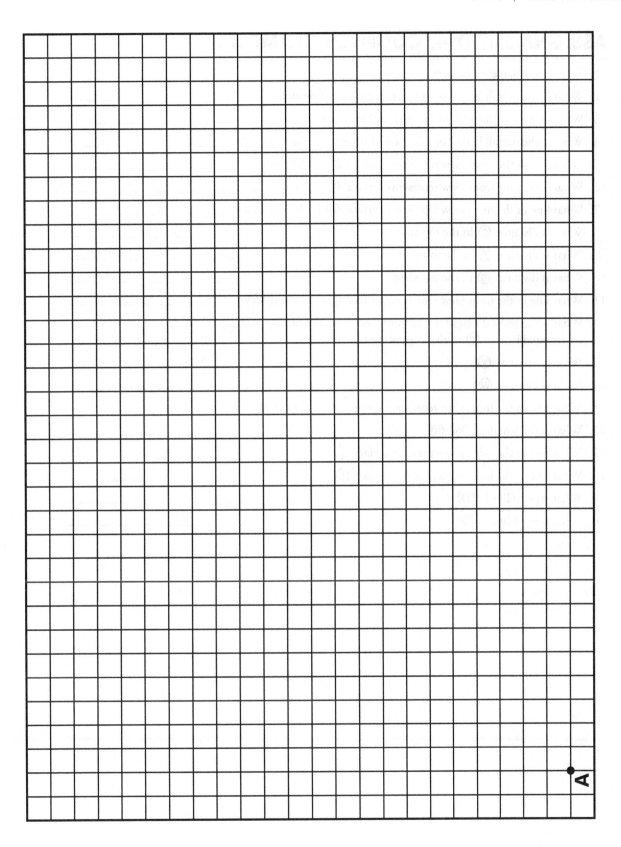

A

ASSIGNMENT D-4: COUNTER CLAMP BAR

1. What is the name of the part? _____

2. What is the overall length of the part in the right-side view? _____

3. What is the overall width of the part in the top view? _____

4. What is the overall height or thickness of the part in the front view? _____

5. What line in the front view represents surface Ⓕ in the top view? _____

6. What line in the front view represents surface Ⓔ in the top view? _____

7. What line in the front view represents surface Ⓖ in the top view? _____

8. What is distance Ⓒ in the top view? _____

9. What is distance Ⓓ in the top view? _____

10. What is distance Ⓑ in the top view? _____

11. What line in the front view represents surface Ⓛ of the side view? _____

12. What is the vertical height in the front view from the surface represented by line Ⓟ to that represented by line Ⓠ? _____

13. What is distance Ⓥ? _____

14. What is distance Ⓦ? _____

15. What line in the front view represents surface Ⓜ in the side view? _____

16. What is the length of line Ⓝ? _____

17. What line in the side view represents surface Ⓡ? _____

18. What line in the top view represents surface Ⓛ? _____

19. What type of line is Ⓣ? _____

20. What type of line is Ⓨ? _____

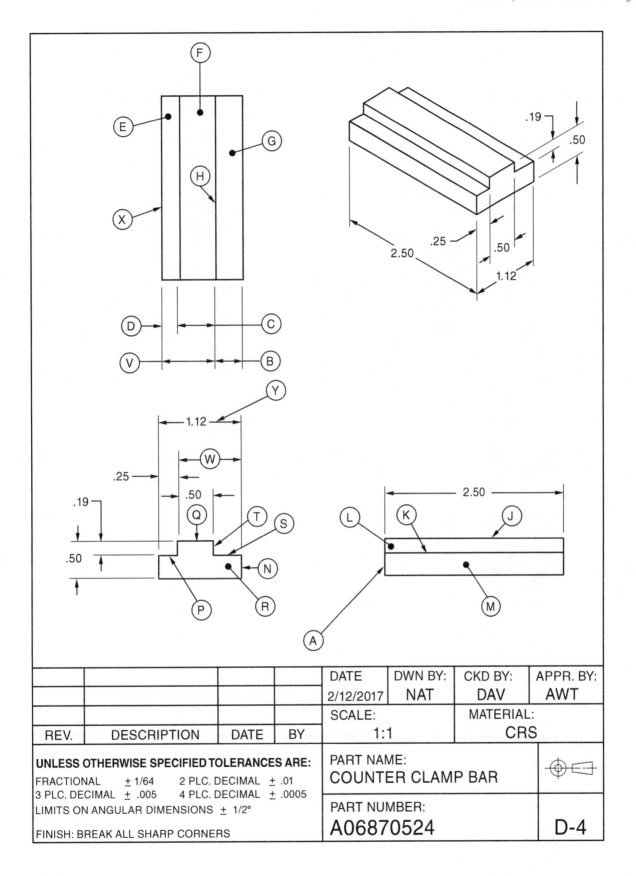

				DATE	DWN BY:	CKD BY:	APPR. BY:
				2/12/2017	NAT	DAV	AWT
				SCALE:		MATERIAL:	
REV.	DESCRIPTION	DATE	BY	1:1		CRS	

UNLESS OTHERWISE SPECIFIED TOLERANCES ARE:

FRACTIONAL ± 1/64   2 PLC. DECIMAL ± .01
3 PLC. DECIMAL ± .005   4 PLC. DECIMAL ± .0005
LIMITS ON ANGULAR DIMENSIONS ± 1/2°

FINISH: BREAK ALL SHARP CORNERS

PART NAME:
COUNTER CLAMP BAR

PART NUMBER:
A06870524

D-4

UNIT 10
TWO-VIEW DRAWINGS

TWO-VIEW DRAWINGS

Some complex cylindrical or flat objects may require two views to provide full understanding of the part. When two views are needed, the drafter must select the views that will best describe the true size and shape of the object in greatest detail. Usually combinations of the front and right-side or front and top views are used. However, other views such as a side view or bottom view may be used as well. The selection is determined by eliminating any unnecessary view. An unnecessary view is one that repeats the shape description of another view. Figures 10.1A and 10.1B show two examples of unnecessary views.

PROJECTING CYLINDRICAL WORK

Cylindrical pieces that may require more than one view include shafts, collars, studs, and bolts. The view chosen for the front shows the length and shape of the object. The top view or side view describes the object as it might be seen looking at the end. A top or side view of the object shows the circular shape of the part.

The cylindrical piece always has a centerline through its axis. In the circular view, a small cross indicates the center of the object, Figure 10.2. The dimension of the diameter is generally given in the same view as the length of the object, Figure 10.3.

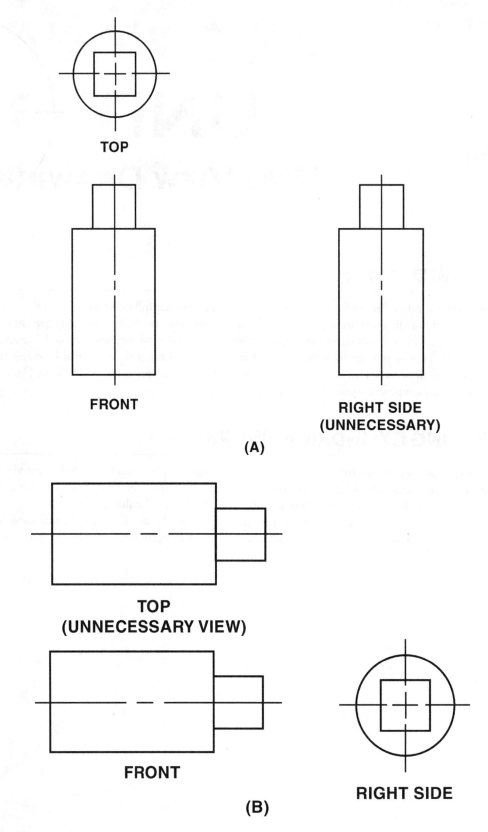

FIGURE 10.1 Unnecessary views.

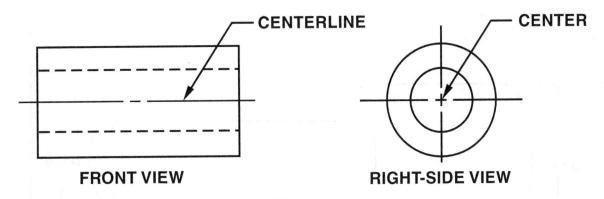

FIGURE 10.2 Cylindrical object shown in two views.

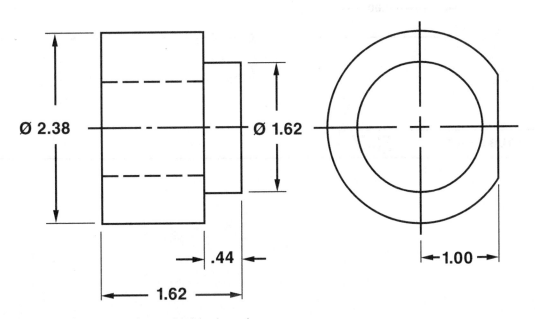

FIGURE 10.3 Dimensioning cylindrical work.

SKETCH S-7: STUB SHAFT

1. Using the centerlines provided in the following grid, sketch front and right-side views of the stub shaft. Allow .75 inch between views.
2. Dimension the completed sketch.

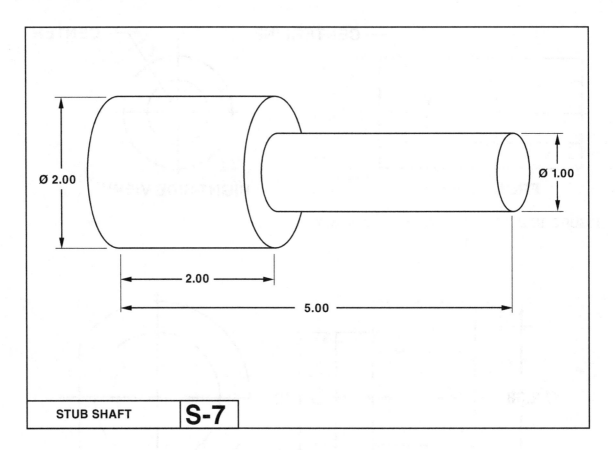

| STUB SHAFT | S-7 |

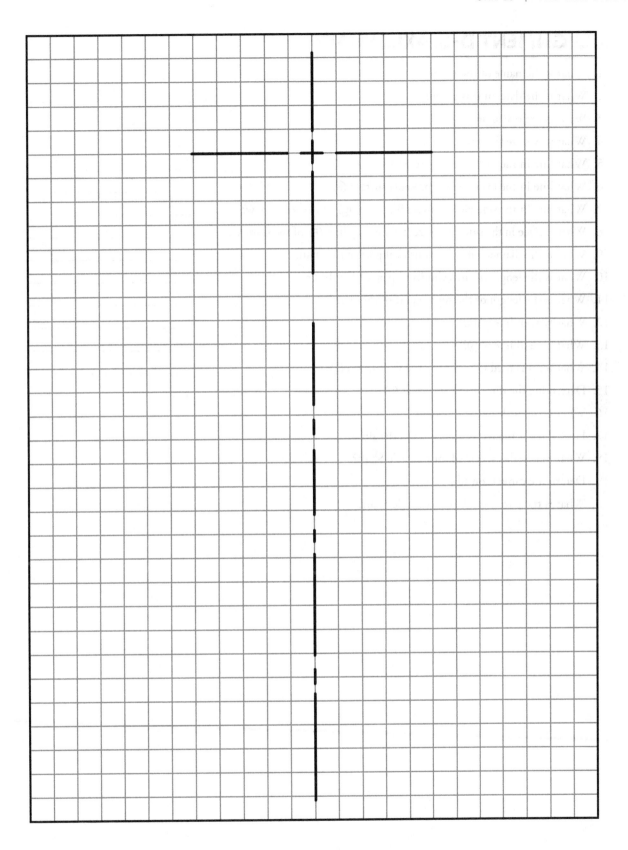

ASSIGNMENT D-5: AXLE SHAFT

1. What is the name of the part?   _____

2. What is the date on this drawing?   _____

3. What is the overall length?   _____

4. What views are shown?   _____

5. What line in the front view represents surface Ⓐ?   _____

6. What line in the front view represents surface Ⓑ?   _____

7. What line in the side view represents surface Ⓒ of the front view?   _____

8. What surface in the side view is represented by line Ⓕ of the front view?   _____

9. What is the diameter of the cylindrical part of the shaft?   _____

10. What is the length of the cylindrical part of the shaft?   _____

11. What is the length of the square part of the shaft?   _____

12. What type of line is Ⓕ?   _____

13. What type of line is Ⓖ?   _____

14. What term is used to designate the dimension Ⓙ?   _____

15. Determine the dimension shown at Ⓗ.   _____

16. What is the scale of the drawing?   _____

17. How thick is the rectangular part of the shaft?   _____

18. What material is specified for the Axle Shaft?   _____

19. Determine dimension Ⓙ.   _____

20. What is the part number for the Axle Shaft?   _____

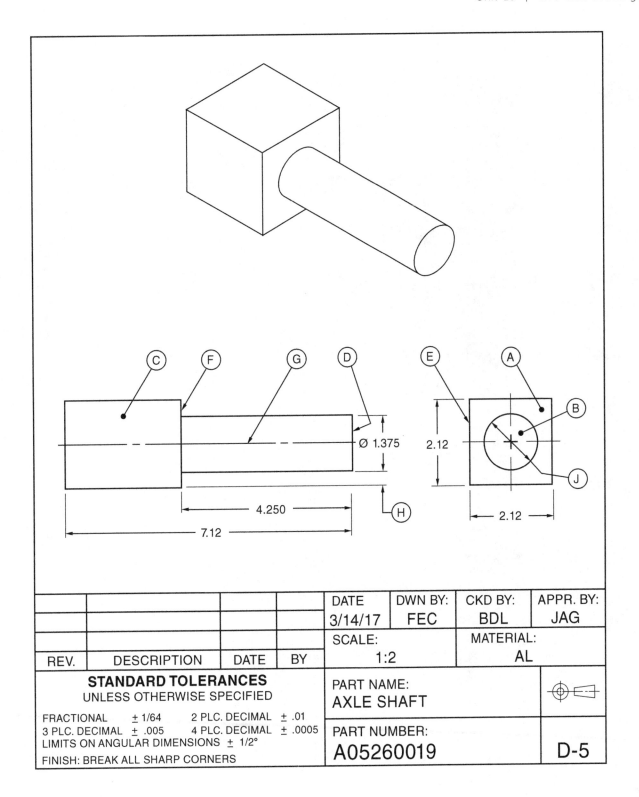

				DATE	DWN BY:	CKD BY:	APPR. BY:
				3/14/17	FEC	BDL	JAG
				SCALE:		MATERIAL:	
REV.	DESCRIPTION	DATE	BY	1:2		AL	
STANDARD TOLERANCES UNLESS OTHERWISE SPECIFIED				PART NAME: AXLE SHAFT			
FRACTIONAL ± 1/64 2 PLC. DECIMAL ± .01 3 PLC. DECIMAL ± .005 4 PLC. DECIMAL ± .0005 LIMITS ON ANGULAR DIMENSIONS ± 1/2° FINISH: BREAK ALL SHARP CORNERS				PART NUMBER: A05260019			D-5

UNIT 11

ONE-VIEW DRAWINGS

ONE-VIEW DRAWINGS

A drawing should show the object in as few views as possible. Simple objects, or objects that are uniform in shape, may not require more than one view for complete size and shape description. This is often the case with cylindrical work such as shafts, pins, or rods. Additional information about the part may be provided in the form of notes or symbols. *One-view drawings* save drafting time and also make the print easier to read because unnecessary views are not shown.

In the case of Figures 11.1 and 11.2, the side or circular views would be omitted. The centerlines and the symbol Ø for the diameter indicate that the objects are cylindrical.

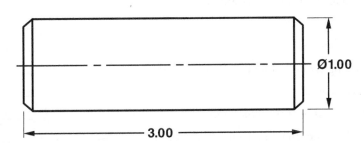

FIGURE 11.1 Only one view is necessary to describe this cylindrical object.

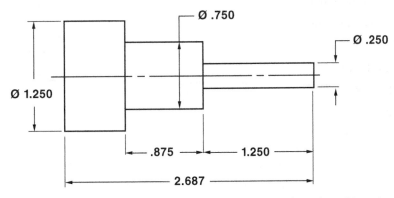

FIGURE 11.2 The centerline and diameter symbols indicate that the object is cylindrical.

Parts that are flat and thin may also be drawn using one view. Notes are added to describe thickness, material, or operations, Figure 11.3. When only one view is drawn, it generally is called a front view.

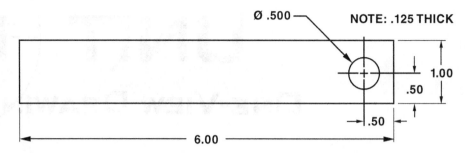

Ø .500

NOTE: .125 THICK

1.00

.50

.50

6.00

FIGURE 11.3   One view and additional notes give all the necessary information about the object.

ASSIGNMENT D-6: SPACER SHIM

1. What is the name of the part?

2. What is the scale of the drawing?

3. What does the 3 X mean?

4. What is the part number?

5. What material is specified for the Spacer Shim?

6. How thick is the part?

7. How many holes are required?

8. What is the overall length of the shim from left to right?

9. What is the distance between holes?

10. What is the radius around the end of the shim?

11. What is the overall height of the shim from top to bottom?

12. What angle forms the top and bottom edges?

13. What size are the holes?

14. What does the Ø symbol indicate?

15. What view is shown in this drawing?

16. What is the date of the drawing?

17. What type of line is Ⓐ?

18. What type of line is Ⓑ?

19. What type of line is Ⓒ?

20. What type of line is Ⓓ?

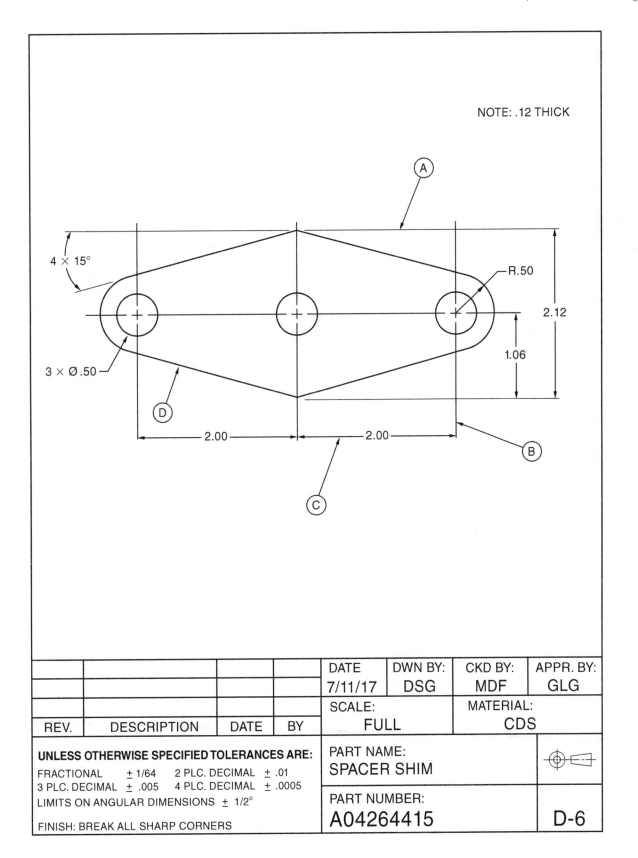

NOTE: .12 THICK

Ⓐ

4 × 15°

R.50

2.12

1.06

3 × Ø.50

Ⓓ

2.00

2.00

Ⓒ

Ⓑ

				DATE	DWN BY:	CKD BY:	APPR. BY:
				7/11/17	DSG	MDF	GLG
				SCALE:		MATERIAL:	
REV.	DESCRIPTION	DATE	BY	FULL		CDS	

UNLESS OTHERWISE SPECIFIED TOLERANCES ARE:

FRACTIONAL ± 1/64   2 PLC. DECIMAL ± .01
3 PLC. DECIMAL ± .005   4 PLC. DECIMAL ± .0005
LIMITS ON ANGULAR DIMENSIONS ± 1/2°

FINISH: BREAK ALL SHARP CORNERS

PART NAME:
SPACER SHIM

PART NUMBER:
A04264415

D-6

UNIT 12

AUXILIARY VIEWS

Many objects have inclined surfaces that will appear distorted in other principal views. Therefore, it may be necessary for the drafter to show the inclined surface in an auxiliary view. An ***auxiliary view*** is an orthographic projection that allows the print reader to see the inclined surface without distortion. For example, in Figure 12.1, surface "x" is shown in an auxiliary view. In the regular top and side views, the circular surfaces are elliptical and the inclined surfaces appear to be shorter. In the auxiliary view, the inclined surfaces are true size and true shape and the hole is round.

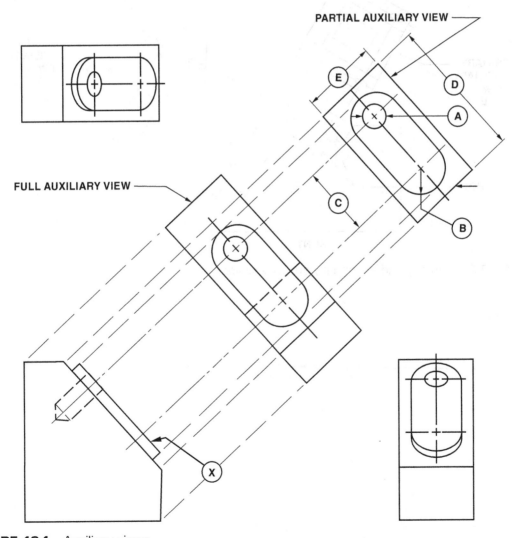

FIGURE 12.1 Auxiliary views.

A *full auxiliary view* shows a full projection of points, lines, and surfaces in the primary view. However, many times a *partial auxiliary view* that only shows the inclined surface will provide all of the information required.

It is accepted practice to show dimensions on the auxiliary view, since this is the only place where they occur in their true size and shape.

A *primary auxiliary view* at view **A** in Figure 12.2 shows the true projection and true shape of face **C** as it is projected directly from the primary front view. Another primary auxiliary view is shown at view **B**. This view is projected from another side of the primary front view. To show the true shape and location of the holes, a *secondary auxiliary view* is provided. A secondary auxiliary view is projected from a primary auxiliary view and is often required when an object has two or more inclined surfaces.

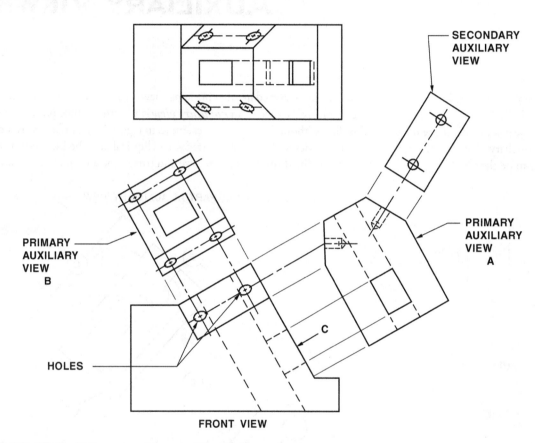

FIGURE 12.2   Primary and secondary auxiliary views.

SKETCH S-8: ANGLE BRACKET

Sketch the top auxiliary view for the angle bracket.

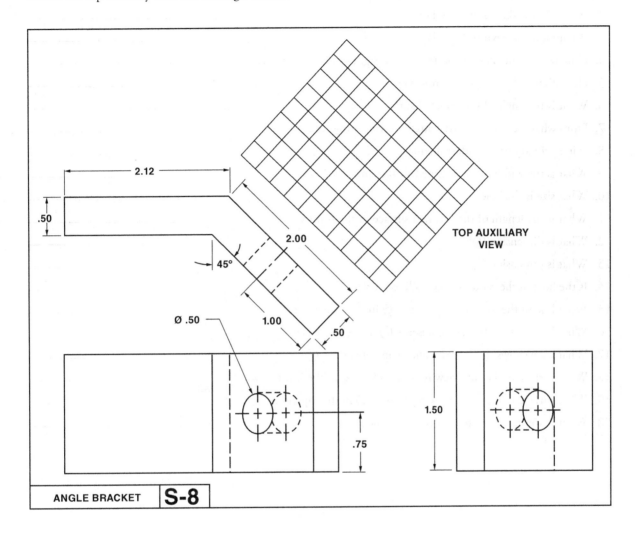

TOP AUXILIARY VIEW

2.12

.50

2.00

45°

1.00

.50

Ø .50

.75

1.50

ANGLE BRACKET **S-8**

ASSIGNMENT D-7: SLIDE BLOCK

1. What view is shown in View I? _____

2. What view is shown in View IV? _____

3. What view is shown in View III? _____

4. How wide is the slide block from left to right in the front view? _____

5. How high is the part in the front view? _____

6. What is the angle of the inclined surface? _____

7. From what view is the partial auxiliary view projected? _____

8. Is it a primary or secondary auxiliary view? _____

9. What is the width of the inclined surface in the side view? _____

10. What size is the hole? _____

11. What is the length of the inclined surface in the front view? _____

12. What is dimension Ⓐ? _____

13. What is dimension Ⓑ? _____

14. Is the hole in the center of the inclined surface? _____

15. What line in the top view represents Ⓓ in the side view? _____

16. What line in the side view represents Ⓖ in the top view? _____

17. What view shows the true size and shape of surface Ⓗ? _____

18. What surface in the side view represents line Ⓔ in the front view? _____

19. What surface in the front view represents Ⓖ in the top view? _____

20. What material is required for the Slide Block? _____

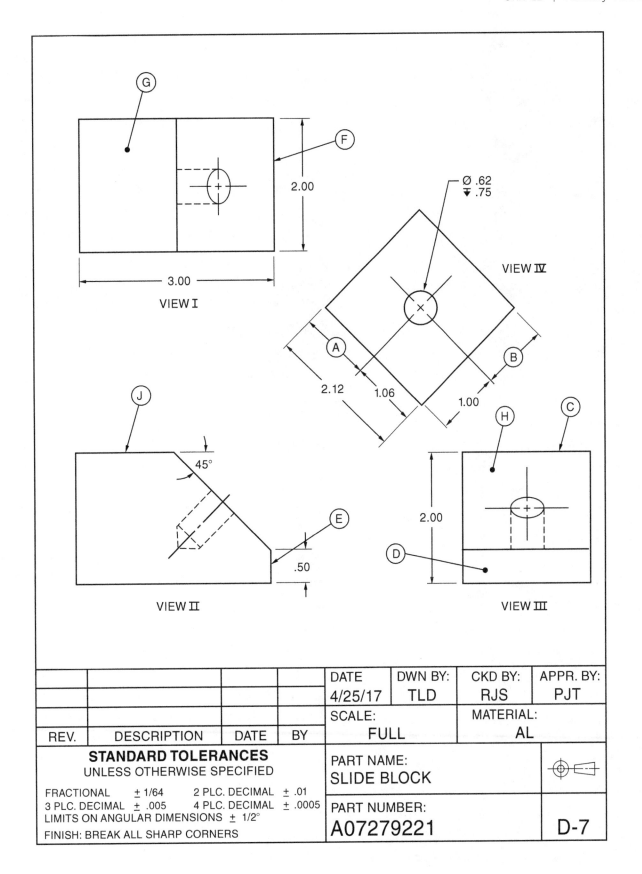

VIEW I

VIEW II

VIEW III

VIEW IV

G

F

2.00

3.00

Ø .62
.75

A

B

2.12

1.06

1.00

J

45°

E

.50

H

C

D

2.00

DATE	DWN BY:	CKD BY:	APPR. BY:
4/25/17	TLD	RJS	PJT

SCALE:	MATERIAL:
FULL	AL

REV.	DESCRIPTION	DATE	BY

STANDARD TOLERANCES
UNLESS OTHERWISE SPECIFIED

FRACTIONAL   ± 1/64   2 PLC. DECIMAL ± .01
3 PLC. DECIMAL ± .005   4 PLC. DECIMAL ± .0005
LIMITS ON ANGULAR DIMENSIONS ± 1/2°
FINISH: BREAK ALL SHARP CORNERS

PART NAME:
SLIDE BLOCK

PART NUMBER:
A07279221

D-7

UNIT 13

SECTION VIEWS

The details of the interior of an object may be shown more clearly if the object is drawn as though a part of it were cut away, exposing the inside surfaces. When showing an object in section, all surfaces that were hidden are drawn as visible surface lines. The surfaces that have been cut through are indicated by a series of slant lines known as *section lining*.

CUTTING PLANE LINE

The line that indicates the plane cutting through the object is the *cutting plane line*, Figure 13.1. After being cut, the portion of the object to the right of the cutting plane in Figure 13.1 is considered to be removed. The portion to the left of the cutting plane is viewed in the direction of the arrows, as shown in section A-A.

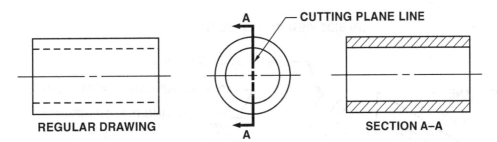

FIGURE 13.1   Sectioning of a hollow cylinder.

FULL SECTIONS

The type and number of sections depend on the complexity of the part. A *full section* is one in which an imaginary cut has been made all the way through the object. The cut section of the object is then represented in a separate view called a *section view*. Hidden lines representing surfaces behind the cutting plane are left out. This helps to keep the view clear for better understanding. When additional section views of a part are shown, they are identified by using different set of letters such as B-B, C-C, etc.

HALF SECTIONS

A half section is often used for symmetrical objects. A *half section* shows a cutaway view of only one half of the part, Figure 13.2 (C). One advantage of half sections is that both an interior and exterior view are shown in the same view.

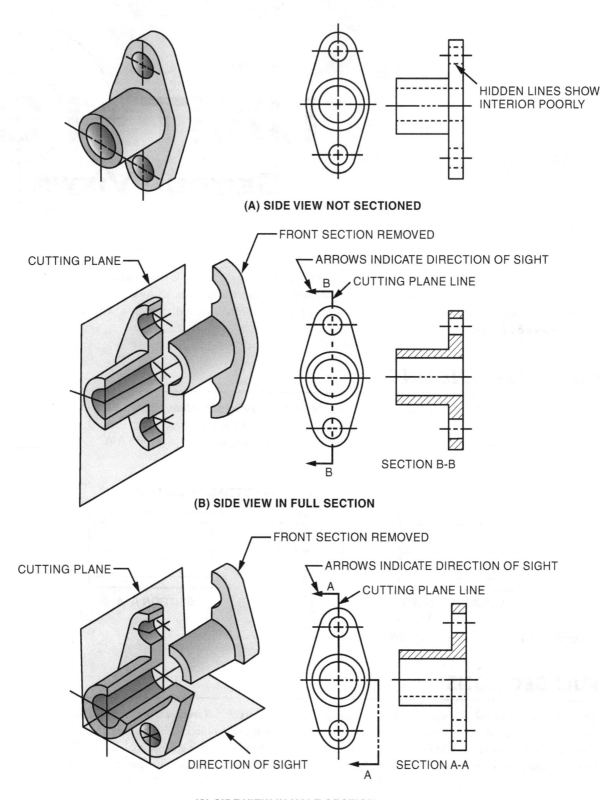

(A) SIDE VIEW NOT SECTIONED

HIDDEN LINES SHOW INTERIOR POORLY

FRONT SECTION REMOVED

ARROWS INDICATE DIRECTION OF SIGHT

CUTTING PLANE

CUTTING PLANE LINE

B

B

SECTION B-B

(B) SIDE VIEW IN FULL SECTION

FRONT SECTION REMOVED

ARROWS INDICATE DIRECTION OF SIGHT

CUTTING PLANE

CUTTING PLANE LINE

A

DIRECTION OF SIGHT

A

SECTION A-A

(C) SIDE VIEW IN HALF SECTION

FIGURE 13.2 Full and half sections.

OFFSET CUTTING PLANES

Offsetting, or changing the straight-line direction of the cutting plane, is used to describe internal features of irregular objects. These features may not fall in a straight line and, therefore, the cutting plane line must be offset to pass through them, Figure 13.3.

Offset cutting plane lines may also be used to show partial sections as in Figure 13.4.

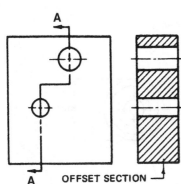

FIGURE 13.3 Offset cutting plane line.

FIGURE 13.4 Offset cutting plane lines showing partial section.

REVOLVED SECTIONS

Revolved sections are used to show the cross-sectional shape of spokes, ribs, cast arms, rods, or structural steel shapes. The view is obtained by passing a cutting plane through the object perpendicular to the centerline or axis of the part. The section is then rotated 90 degrees on the view to reflect the true size and shape of the area in the section. Visible lines on either side of the section are removed or break lines are added to isolate the view, Figure 13.5. Cutting plane lines are usually omitted in revolved section views.

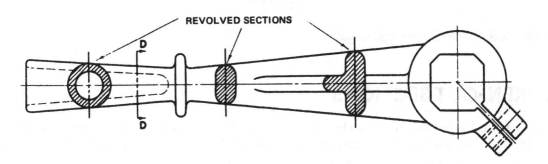

FIGURE 13.5 Revolved sections.

REMOVED SECTIONS

Removed sections are very similar to revolved sections. However, a **removed section** is taken out of the normal projection position in relation to standard views. A removed section is usually placed on the drawing in a convenient place and labeled section A-A, section B-B, and so on. The letters identifying the section correspond with the letters at the end of the cutting plane lines, Figure 13.6.

A removed section often is a partial section view. Frequently, the view is enlarged to permit greater clarification, Figure 13.7 and Figure 13.8.

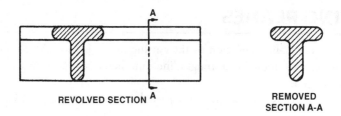

FIGURE 13.6 Removed section.

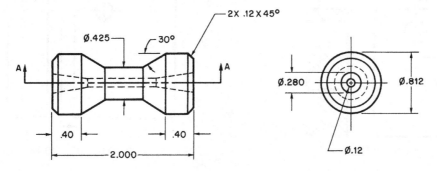

FIGURE 13.7 Full section A-A.

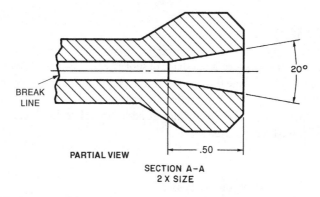

FIGURE 13.8 Labeled removed section.

BROKEN-OUT SECTIONS

When only a partial section view is needed, a broken-out section may be used. A ***broken-out section*** isolates one area of the object for internal clarification. A heavy break line is used to define the boundaries of the view, Figure 13.9. Broken-out sections are used where less than a half section is needed.

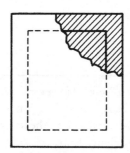

FIGURE 13.9 Broken-out section.

MATERIALS IN SECTION

In drawing sections of various machine parts, section lines indicate the different materials of which the parts are made, Figure 13.10. Each material is represented by a different pattern of lines. On most drawings, however, sections are shown using the pattern for cast iron. The kind of material is then indicated in the specifications.

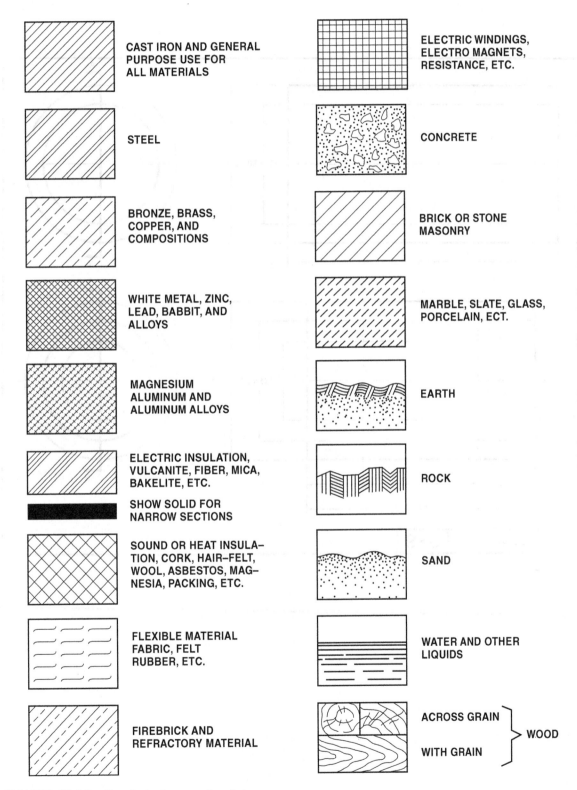

CAST IRON AND GENERAL PURPOSE USE FOR ALL MATERIALS

STEEL

BRONZE, BRASS, COPPER, AND COMPOSITIONS

WHITE METAL, ZINC, LEAD, BABBIT, AND ALLOYS

MAGNESIUM ALUMINUM AND ALUMINUM ALLOYS

ELECTRIC INSULATION, VULCANITE, FIBER, MICA, BAKELITE, ETC.

SHOW SOLID FOR NARROW SECTIONS

SOUND OR HEAT INSULATION, CORK, HAIR–FELT, WOOL, ASBESTOS, MAGNESIA, PACKING, ETC.

FLEXIBLE MATERIAL FABRIC, FELT RUBBER, ETC.

FIREBRICK AND REFRACTORY MATERIAL

ELECTRIC WINDINGS, ELECTRO MAGNETS, RESISTANCE, ETC.

CONCRETE

BRICK OR STONE MASONRY

MARBLE, SLATE, GLASS, PORCELAIN, ECT.

EARTH

ROCK

SAND

WATER AND OTHER LIQUIDS

ACROSS GRAIN } WOOD
WITH GRAIN

FIGURE 13.10 Symbols for section lining.

SKETCH S-9: COLLARS

1. Copy the drawing of the collars on the grid.

2. Cut through sections A-A and B-B as indicated. Show the front in section.

3. Sketch in section lining.

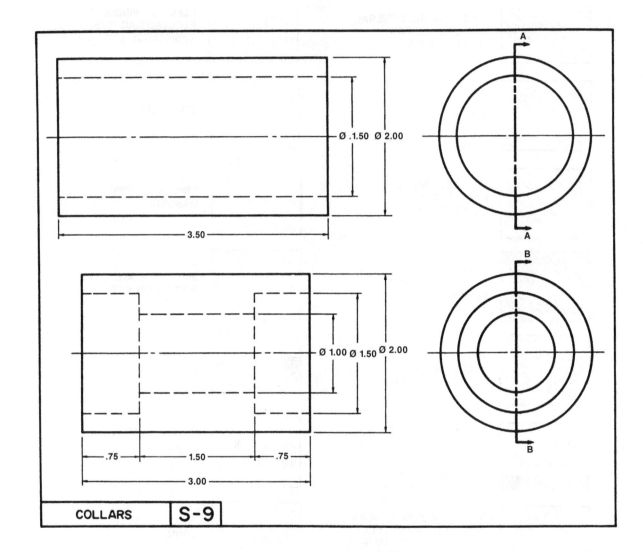

ASSIGNMENT D-8: TOOL POST BLOCK

1. What heat treatment is specified in the note on the drawing? _____

2. What is the scale of the drawing? _____

3. What is the length and the width of the Tool Post Block? _____

4. How thick is it? _____

5. What type of section is shown at A–A? _____

6. What type of line is shown at Ⓔ? _____

7. What do the arrows at the end of the line Ⓔ indicate? _____

8. What surface in the top view does line Ⓓ represent? _____

9. What surface in the top view does line Ⓒ represent? _____

10. What is the diameter of the smaller hole in the block? _____

11. How deep is the smaller hole? _____

12. How thick is the material between surface Ⓓ and surface Ⓕ? _____

13. What finish is specified in the title block? _____

14. What material does the section lining indicate? _____

15. What radius is required in the bottom of the Ø 2.000 hole? _____

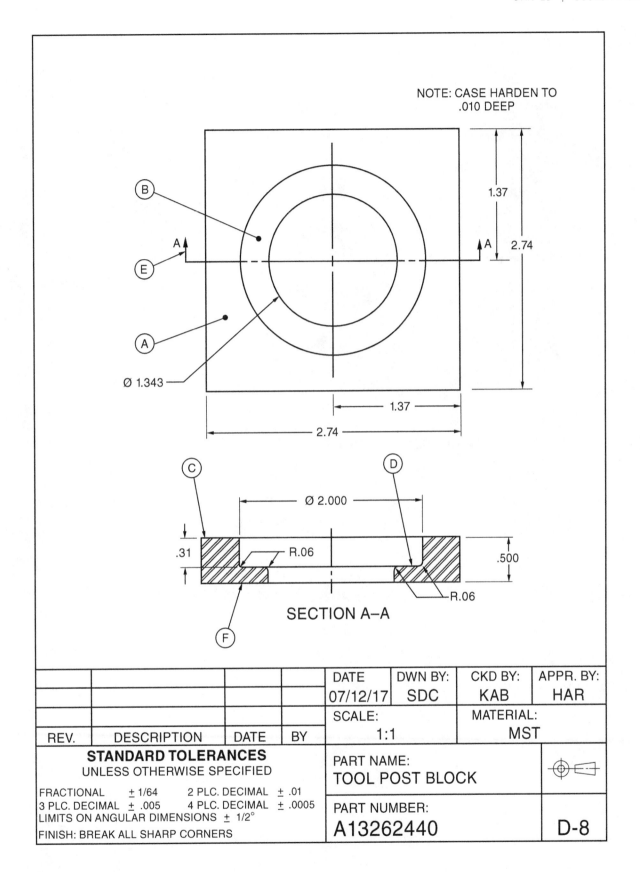

NOTE: CASE HARDEN TO
.010 DEEP

B

A

E

A

Ø 1.343

1.37

2.74

1.37

2.74

C

D

Ø 2.000

.31

R.06

.500

R.06

SECTION A–A

F

				DATE	DWN BY:	CKD BY:	APPR. BY:
				07/12/17	SDC	KAB	HAR
				SCALE:		MATERIAL:	
REV.	DESCRIPTION	DATE	BY	1:1		MST	

STANDARD TOLERANCES	PART NAME:	
UNLESS OTHERWISE SPECIFIED	TOOL POST BLOCK	
FRACTIONAL ± 1/64 2 PLC. DECIMAL ± .01 3 PLC. DECIMAL ± .005 4 PLC. DECIMAL ± .0005 LIMITS ON ANGULAR DIMENSIONS ± 1/2° FINISH: BREAK ALL SHARP CORNERS	PART NUMBER: A13262440	D-8

UNIT 14

DIMENSIONS AND TOLERANCES

An industrial drawing should provide all required information about the size and shape of an object. The print reader must be able to visualize the completed part described on the drawing. In previous units, various views that are used to show the shape of an object have been explained. However, a complete size description is also needed to understand what the machining requirements are.

The size of an object is shown by placing measurements, called *dimensions*, on the drawing. Each dimension has limits of accuracy within which it must fall. These limits are called *tolerances*.

In the following units, the types of dimensions and tolerances used on industrial drawings are discussed.

DIMENSIONS

The size requirements on a drawing may be given in any one or a combination of measuring systems. Dimensions may be fractional, decimal, or angular. Each system will be discussed in detail in later units.

As explained in Unit 3, Lines and Symbols, there are special types of lines used in dimensioning. They are called extension lines, dimension lines, and leader lines. Each has a specific purpose as it is applied to the drawing.

In industrial practice there are a few rules followed in dimensioning a drawing. Understanding these rules is helpful in drawing interpretations and shop sketching. The most common rules are:

- Drawings should supply only those dimensions required to produce their intended objects.

- Dimensions should not be duplicated on a drawing. If a dimension is provided in one view, it should not be given in the other views. Duplicate or double dimensioning is redundant, permits error, and can lead to confusion in print interpretation.

- Dimensions should be placed between views where possible. This helps in identifying points and surface dimensions in adjacent views.

- Dimensions should be spaced from the outside of the object in order of size, Figure 14.1. Smaller dimensions are placed closer to the parts they dimension.

- Notes should be added to dimensions where drawing clarification is required.

- Dimensions should not be placed on the view, if possible.

- Hidden surfaces should not be dimensioned, if possible.

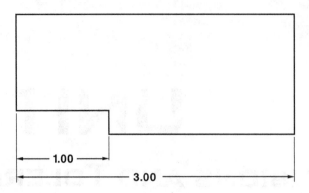

FIGURE 14.1 Dimensions are spaced in order of size.

Types of Dimensions

Dimensions placed on drawings are identified as either size or location dimensions. *Size dimensions* are used to indicate lengths, widths, or thicknesses, Figure 14.2. *Location dimensions* are used to show the location of holes, points, or surfaces, Figure 14.3. Both types are often called *construction dimensions*.

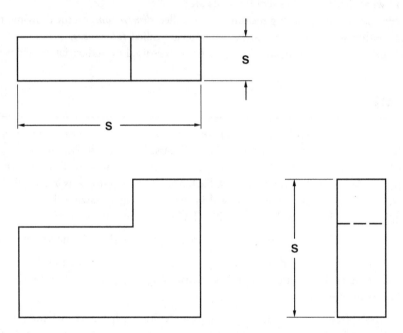

FIGURE 14.2 Size dimensions.

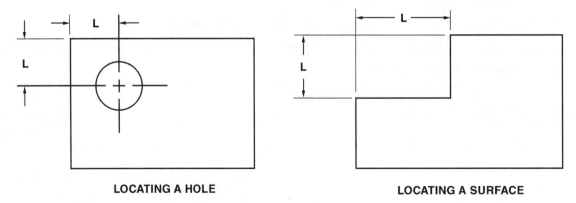

LOCATING A HOLE

LOCATING A SURFACE

FIGURE 14.3 Location dimensions.

Methods of Dimensioning

There are two common systems of dimensioning used on industrial drawings. The *aligned method* is read from the bottom and right side of the drawing. Reading dimensions done in this system often requires turning the drawing. The aligned method is still used, but it is being replaced with a second system called unidirectional.

The *unidirectional dimensions* are all read from the bottom. Therefore, it is not necessary to turn the print. Figure 14.4 shows the unidirectional and aligned systems of dimensioning.

Note: The most recent ASME standards recommend the use of unidirectional dimensioning. The drawings in this text, therefore, are all shown with unidirectional dimensions.

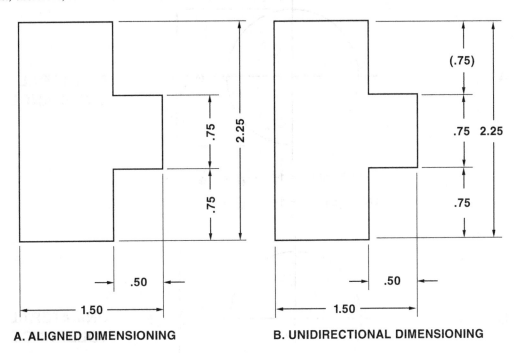

FIGURE 14.4 Methods of dimensioning.

TOLERANCES

Because it is nearly impossible to make anything to exact size, degrees of accuracy must be specified. When a size is given on a drawing, a tolerance is applied to it. The *tolerance* is a range of sizes within which the actual dimension of a piece must fall. The tolerance specifies how exact the dimension must be.

Just as in dimensioning, the tolerances may be fractional, decimal, or angular. Tolerances may be given in the title block area, Figure 14.5, or on the dimension itself. Tolerances given in the title block apply to all dimensions unless otherwise specified on the drawing.

STANDARD TOLERANCE UNLESS OTHERWISE SPECIFIED			
MILLIMETER		INCH	
WHOLE No.	± .5	FRACTION	± 1/164
1 PL DEC.	± .2	2 PL DEC.	± .008
2 PL DEC.	± .03	3 PL DEC.	± .001
3 PL DEC.	± .013	4 PL DEC.	± .0005
ANGLES ± .5 DEGREES			

FIGURE 14.5 Dimensional tolerance block.

Upper and Lower Limits

All dimensions to which a tolerance is applied have upper and lower limits of size. The ***upper limit*** is the print dimension with the (+) tolerance added to it. If no (+) tolerance is allowed, the print dimension becomes the upper limit.

The ***lower limit*** dimension is the print dimension with the (–) tolerance subtracted. If no (–) tolerance is allowed, then the print dimension becomes the lower limit, Figure 14.6.

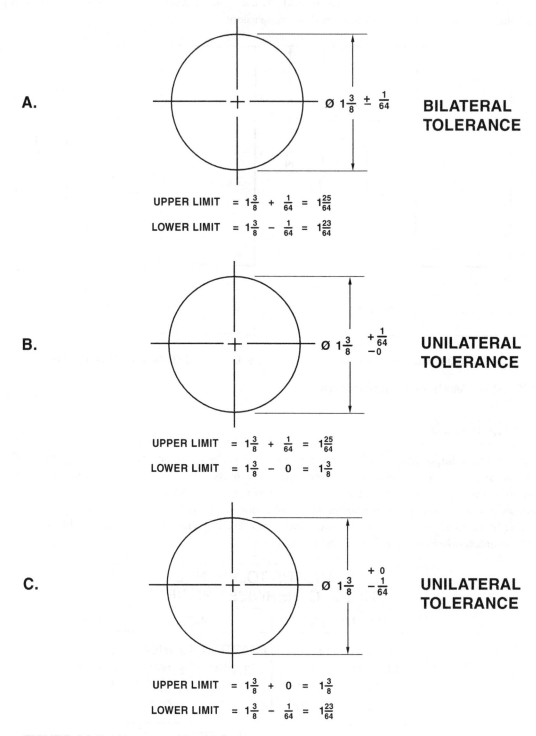

A.

$\emptyset \; 1\frac{3}{8} \pm \frac{1}{64}$ **BILATERAL TOLERANCE**

UPPER LIMIT $= 1\frac{3}{8} + \frac{1}{64} = 1\frac{25}{64}$

LOWER LIMIT $= 1\frac{3}{8} - \frac{1}{64} = 1\frac{23}{64}$

B.

$\emptyset \; 1\frac{3}{8} \; {}^{+\frac{1}{64}}_{-0}$ **UNILATERAL TOLERANCE**

UPPER LIMIT $= 1\frac{3}{8} + \frac{1}{64} = 1\frac{25}{64}$

LOWER LIMIT $= 1\frac{3}{8} - 0 = 1\frac{3}{8}$

C.

$\emptyset \; 1\frac{3}{8} \; {}^{+0}_{-\frac{1}{64}}$ **UNILATERAL TOLERANCE**

UPPER LIMIT $= 1\frac{3}{8} + 0 = 1\frac{3}{8}$

LOWER LIMIT $= 1\frac{3}{8} - \frac{1}{64} = 1\frac{23}{64}$

FIGURE 14.6 Upper and lower limits.

Methods of Tolerancing

The two systems of tolerancing are known as bilateral and unilateral tolerances. A ***bilateral tolerance*** allows for variation in two directions from the print dimension. A tolerance is given as both a plus (+) and a minus (–) dimension. For example, 1 3/8 ± 1/64, Figure 14.6A. Bilateral tolerances may not always be an equal amount in each direction. A ***unilateral tolerance*** allows for variation in only one direction from the print dimension. The tolerance may be a (+) or a (–) dimension from the print dimension. For example, 1 3/8 + 1/64 or 1 3/8 – 1/64, Figures 14.6B and 14.6C.

FRACTIONAL DIMENSIONS

Perhaps the oldest dimensioning system used is ***fractional*** dimensioning. This system divides an inch unit into fractional parts of an inch, with 1/64 being the smallest fraction used. Decimal dimensions have largely replaced the fractional system due to modern requirements for accuracy, close tolerances, and the use of precision measuring and inspection tools. However, fractions are still found on older part prints, assemblies, fabricated components, piping, tooling fixtures, standard material sizes, and cutting tools such as drills and reamers.

It is important for the print reader to have a basic understanding of this system of measurement. Assignment D-9, Idler Shaft, is provided in this unit to provide practice working with fractions.

Fractional dimensioning is used where close tolerances are not required. This is often the case on castings, forgings, standard material sizes, bolts, drilled holes, or machine parts where exact size is unimportant.

FRACTIONAL TOLERANCES

Fractional dimensions usually have a ***fractional tolerance*** applied to them. As shown in Figure 14.6, this tolerance may be unilateral or bilateral.

ASSIGNMENT D-9: IDLER SHAFT

1. How many hidden diameters are shown in the top view of the idler shaft?   _____

2. What is the diameter shown by the hidden line?   _____

3. What is the diameter of the smallest visible circle?   _____

4. What is the diameter of the largest visible circle?   _____

5. What is the length of that portion which is Ø 1 1/4?   _____

6. What is the length of that portion which is Ø 2 3/4?   _____

7. What is the length of that portion which is Ø 7/8?   _____

8. What is the limit of tolerance on fractional dimensions?   _____

9. What is the largest size the Ø 2 3/4 can be machined?   _____

10. What is the smallest size the Ø 2 3/4 can be machined?   _____

11. If the length of that portion which is Ø 1 1/4 is machined to the upper limit, how long will it be?   _____

12. If the length of the Ø 1 1/4 portion is machined to 2 1/8 inches, how much over the upper limit of size for this length will it be?   _____

13. How much is it below the lower limit if it is 2 1/16 long?   _____

14. How long is the shaft from surface Ⓐ to surface Ⓑ?   _____

15. If the Ø 2 3/4 measures 2 25/32, will it be over, under, or within the limits of accuracy?   _____

16. How much over the upper limit of size will the Ø 2 3/4 be, if it is 2 27/32?   _____

17. What two views of the idler shaft are shown?   _____

18. Could the idler shaft have been shown in one view?   _____

19. What other two views could have been used?   _____

20. What would the overall length of the shaft be if made to the upper limit?   _____

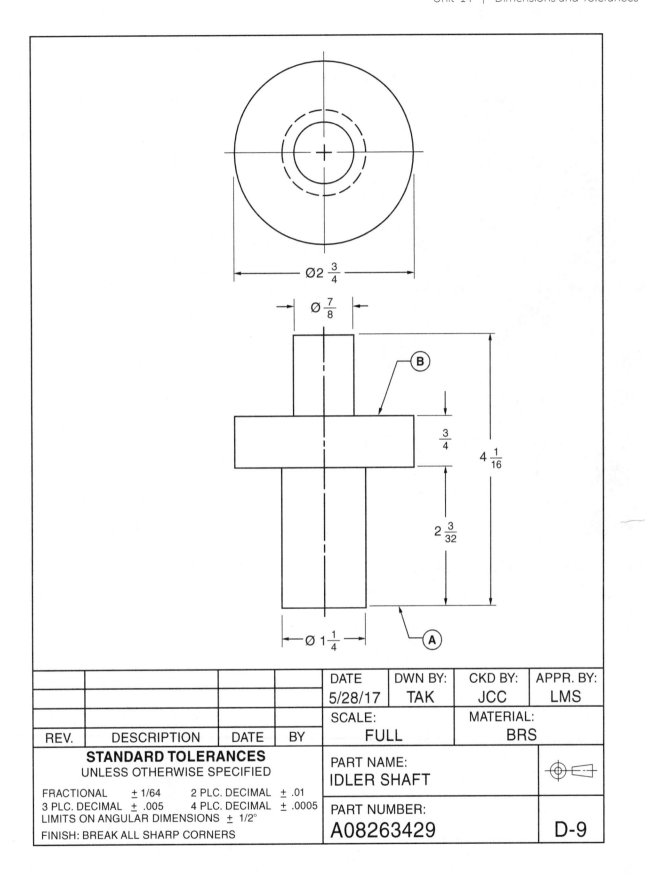

				DATE	DWN BY:	CKD BY:	APPR. BY:
				5/28/17	TAK	JCC	LMS
				SCALE:		MATERIAL:	
REV.	DESCRIPTION	DATE	BY	FULL		BRS	

STANDARD TOLERANCES	PART NAME:	
UNLESS OTHERWISE SPECIFIED	IDLER SHAFT	
FRACTIONAL ± 1/64 2 PLC. DECIMAL ± .01 3 PLC. DECIMAL ± .005 4 PLC. DECIMAL ± .0005 LIMITS ON ANGULAR DIMENSIONS ± 1/2° FINISH: BREAK ALL SHARP CORNERS	PART NUMBER: A08263429	D-9

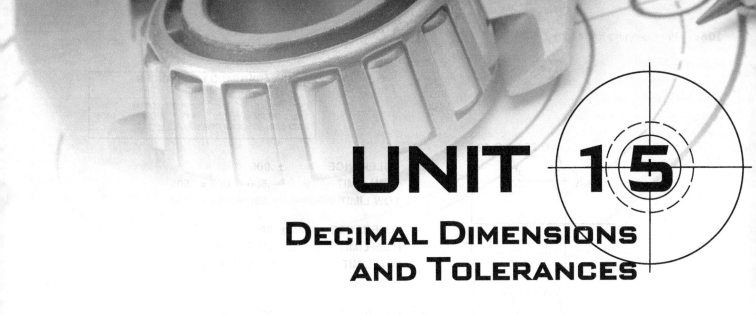

UNIT 15
DECIMAL DIMENSIONS AND TOLERANCES

DECIMAL DIMENSIONS

The need for increased accuracy and closer tolerances in machining led to the development of the decimal system. Today the decimal system of measurement has all but replaced the fractional system when high accuracy is desired. The most common decimal units found on industrial drawings are tenths, hundredths, thousandths, ten-thousandths, and hundred-thousandths. For example:

- One tenth $= 1/10$ $= .10$
- One hundredth $= 1/100$ $= .01$
- One thousandth $= 1/1000$ $= .001$
- One ten-thousandth $= 1/10,000$ $= .0001$

The unit used depends on the degree of accuracy required for the part. The dimension specified must take into consideration the machining process used. The decimal units on industrial drawings seldom exceed four places for dimensioning. This is due to the fact that machine tools and measuring instruments are usually only accurate to three or four decimal places.

Industrial drawings may be all decimal dimensions, or a combination of both decimal and fractional dimensions. The trend, however, has been to dimension totally in decimals to avoid the confusion of using both systems. Decimal dimensions are preferred because they are easier to work with. They may be added, subtracted, divided, and multiplied with fewer problems in calculation. Decimal numbers may also be directly applied to shop measuring tools, machine tool graduation, modern digital readouts, and computer plots.

DECIMAL TOLERANCES

Just as there are tolerances on fractional dimensions, there are tolerances on decimal dimensions. Decimals in thousandths of an inch are used when greater precision and less tolerance is required to make a part.

Decimal tolerances may be specified in the tolerance block or on a drawing in various ways, Figure 15.1.

POINT-TO-POINT DIMENSIONS

Most linear (in line) dimensions apply on a point-to-point basis. *Point-to-point dimensions* are applied directly from one feature to another, Figure 15.2. Such dimensions are intended to locate surfaces and features directly between the points indicated. They also locate corresponding points on the indicated surfaces.

For example, a diameter applies to all diameters throughout the length of a cylindrical surface. It does not merely apply to the diameter at the end where the dimension is shown. A thickness applies to all opposing points on the surfaces.

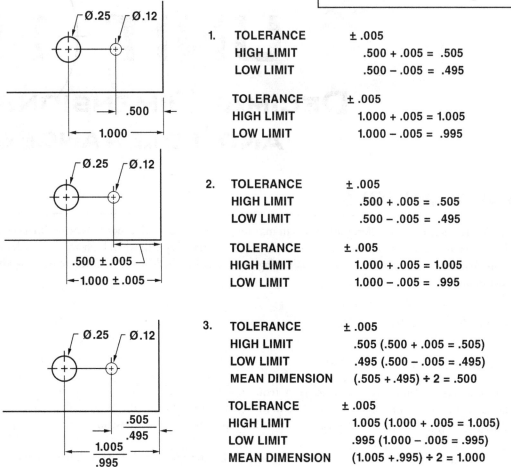

STANDARD TOLERANCES		
UNLESS OTHERWISE SPECIFIED		
FRACTIONAL ± 1/64	2 PLC. DECIMAL ± 0.01	
3 PLC. DECIMAL ± .005	4 PLC. DECIMAL ± .0005	
LIMITS ON ANGULAR DIMENSIONS ± 1/2°		

1. TOLERANCE   ± .005
    HIGH LIMIT   .500 + .005 = .505
    LOW LIMIT   .500 − .005 = .495

    TOLERANCE   ± .005
    HIGH LIMIT   1.000 + .005 = 1.005
    LOW LIMIT   1.000 − .005 = .995

2. TOLERANCE   ± .005
    HIGH LIMIT   .500 + .005 = .505
    LOW LIMIT   .500 − .005 = .495

    TOLERANCE   ± .005
    HIGH LIMIT   1.000 + .005 = 1.005
    LOW LIMIT   1.000 − .005 = .995

3. TOLERANCE   ± .005
    HIGH LIMIT   .505 (.500 + .005 = .505)
    LOW LIMIT   .495 (.500 − .005 = .495)
    MEAN DIMENSION   (.505 + .495) ÷ 2 = .500

    TOLERANCE   ± .005
    HIGH LIMIT   1.005 (1.000 + .005 = 1.005)
    LOW LIMIT   .995 (1.000 − .005 = .995)
    MEAN DIMENSION   (1.005 + .995) ÷ 2 = 1.000

FIGURE 15.1   Specifying decimal tolerances on a drawing.

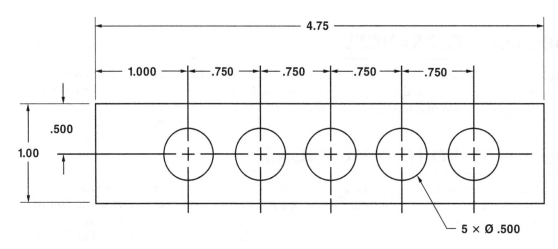

FIGURE 15.2   Point-to-point dimensioning.

RECTANGULAR COORDINATE DIMENSIONING

Rectangular coordinate dimensioning, often called datum or baseline dimensioning, is a system where dimensions are given from one or more common data points. Linear dimensions are typically specified from two or three perpendicular planes, Figure 15.3.

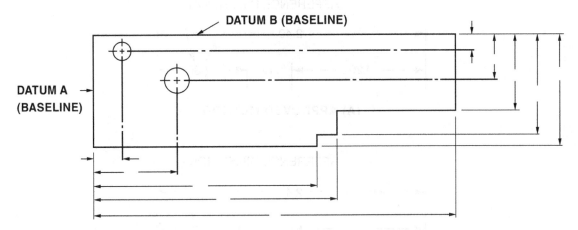

FIGURE 15.3 Rectangular coordinate dimensioning.

Rectangular coordinate dimensioning is often used when accurate part layout is required. Having common data points helps overcome errors that may accumulate in the build-up of tolerances in between point-to-point dimensions.

RECTANGULAR COORDINATE DIMENSIONING WITHOUT DIMENSION LINES

Dimensions may also be given without the use of dimension lines. The datum is often shown as a zero line and the dimensions are placed on the extension lines, Figure 15.4.

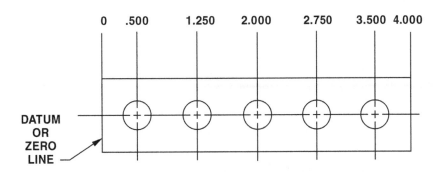

FIGURE 15.4 Rectangular coordinate dimensioning without dimension lines.

Reference Dimensions

Reference (calculated) dimensions may be given for information purposes. The old method was to add the letters REF following the dimension to identify it as a reference dimension, Figure 15.5A. New standards require the reference dimension to be placed within parentheses on the drawing, thus eliminating the REF notation, Figure 15.5B.

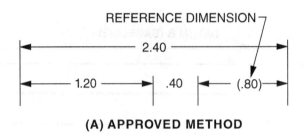

(A) APPROVED METHOD

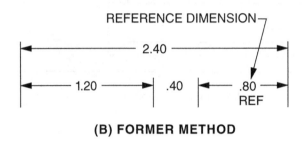

(B) FORMER METHOD

FIGURE 15.5 Reference dimensions.

Dimensions Not to Scale

Objects on original drawings or prints should always be shown true size when possible. There are times, however, when a designer or drafter must insert a dimension or show a surface that is not true size. If a dimension that is not to scale is shown, a straight line is drawn below the dimension, Figure 15.6A.

The old method for showing a dimension that is not to scale was to place a wavy line under the dimension, or add the letters NTS as an abbreviation for Not to Scale, Figure 15.6B.

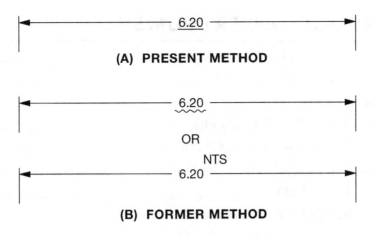

FIGURE 15.6 Dimensions not to scale.

ASSIGNMENT D-10: LOWER DRUM SHAFT

1. What is the overall length of the Lower Drum Shaft? _____

2. What surface is used as the baseline or datum? _____

3. How many diameters are there? _____

4. What is the distance from surface Ⓒ to surface Ⓓ? _____

5. What is the minimum size allowed on the 2.875 diameter? _____

6. What is the maximum size allowed on the 1.250 diameter? _____

7. What is the length of the Ø 1.250? _____

8. Determine the length of the 1.000/.998 diameter? _____

9. Determine the distance from surface Ⓐ to surface Ⓒ? _____

10. Determine the distance from surface Ⓔ to surface Ⓒ? _____

11. Determine the distance from surface Ⓑ to surface Ⓒ. _____

12. How many hidden lines would be drawn in a top view of the part? _____

13. What tolerance is permitted on two-place decimal dimensions where tolerance is not specified? _____

14. What tolerance is permitted on three-place decimal dimensions where tolerance is not specified? _____

15. How many diameters are being held within limits of accuracy smaller than ± .005. _____

16. What is the lower limit of the 2.375 length? _____

17. What is the upper limit of size for the overall length of the shaft? _____

18. What is the lower limit of size for the overall length of the shaft? _____

19. What does the (.38) dimension indicate? _____

20. What does the line under the 1.750 dimension indicate? _____

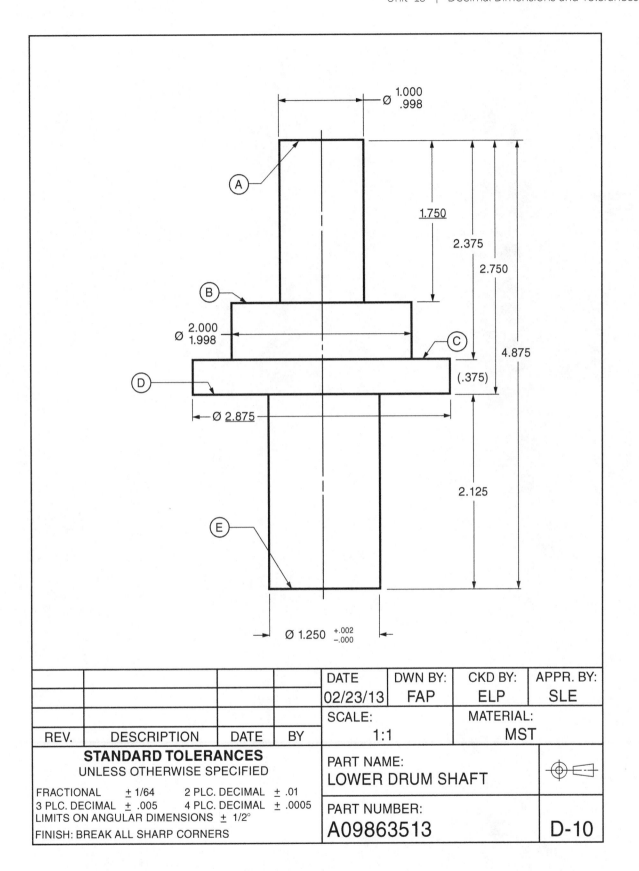

Ø 1.000
.998

A

1.750

2.375

2.750

B

Ø 2.000
1.998

C

4.875

D

(.375)

Ø 2.875

2.125

E

Ø 1.250 +.002
−.000

				DATE	DWN BY:	CKD BY:	APPR. BY:
				02/23/13	FAP	ELP	SLE
				SCALE:		MATERIAL:	
REV.	DESCRIPTION	DATE	BY	1:1		MST	

STANDARD TOLERANCES	PART NAME:	
UNLESS OTHERWISE SPECIFIED	LOWER DRUM SHAFT	
FRACTIONAL ± 1/64 2 PLC. DECIMAL ± .01		
3 PLC. DECIMAL ± .005 4 PLC. DECIMAL ± .0005	PART NUMBER:	
LIMITS ON ANGULAR DIMENSIONS ± 1/2°	A09863513	D-10
FINISH: BREAK ALL SHARP CORNERS		

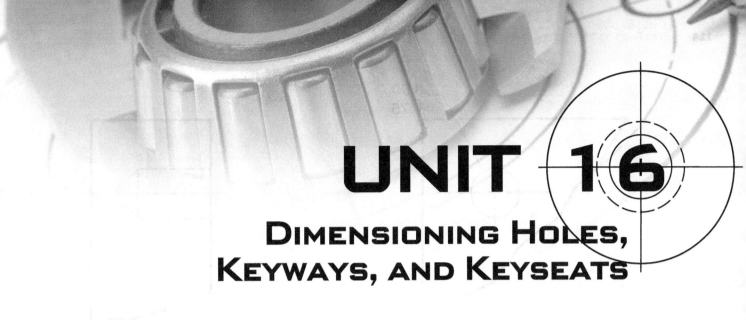

UNIT 16

DIMENSIONING HOLES, KEYWAYS, AND KEYSEATS

DIAMETERS OF HOLES

Holes in objects may be dimensioned in several ways. The method used often depends on the size of the hole or how it is to be produced. Common dimensioning practice is to dimension holes on the view in which they appear as circles. Hole diameters should not be dimensioned on the view in which they appear as hidden lines.

Small hole sizes are shown with a leader line. The leader touches the outside diameter of the hole and points to the center. The hole diameter is given at the end of the leader outside the view of the object.

DEPTH OF HOLES

Holes that go through a part may not require any additional information other than a diameter or repetitive feature callout, Figure 16.1. When it is not clear that a hole goes through the part, however, the letters THRU may follow the hole diameter dimension specified.

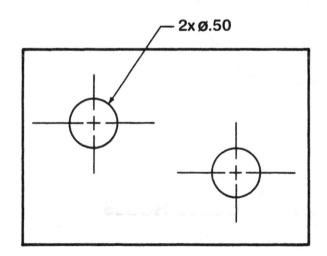
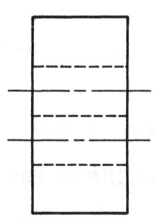

FIGURE 16.1 Specifying hole diameter.

Holes that do not go through a part are commonly referred to as blind holes and must have a note or symbol specifying the depth. The depth of a hole is defined as the length of the full diameter of the hole. It is not the depth of the hole from the outer surface to the point of the drill, for example. In the past, a note was added to the diameter specification indicating the depth of the hole. Current ASME dimensioning standards require the use of a depth symbol, Figure 16.2.

Large holes may be dimensioned within the diameter of the circle on the view, Figure 16.3.

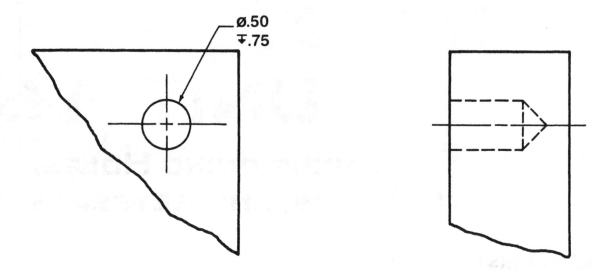

FIGURE 16.2 Specifying hole diameter and depth.

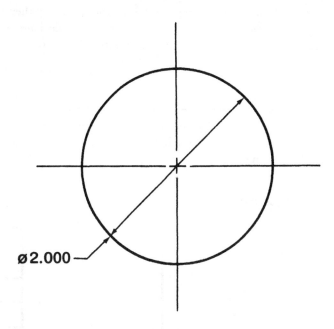

FIGURE 16.3 Dimensioning large diameter holes.

MACHINING PROCESSES FOR PRODUCING HOLES

The former practice was to specify the method for producing a hole on the industrial drawing. The current practice is to specify the hole size and location and allow the manufacturer to determine the process for machining the hole based on the tolerances for size, location, and finish.

Drilling

Drilling is one of the most efficient processes for producing holes in an object. The tool most often used in drilling is called a twist drill. The twist drill gets its name from the spiral flutes along the body of the tool. These flutes enable the drill to carry chips out of the hole. Twist drills come in a variety of standard sizes.

Although drilling is a very efficient operation, it is seldom used when close tolerances and a smooth hole is required. Therefore, drilling is performed most often when close tolerances are not specified, Figure 16.4.

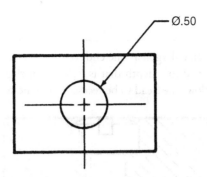

FIGURE 16.4 Drilled hole.

Reaming

Reaming is the process of sizing a hole to a given diameter with a tool called a reamer. Just as in the case of drills, reamers are available in a variety of diameters. Reaming produces a round, straight, smooth hole to close tolerances, Figure 16.5.

Holes that are to be reamed are first drilled slightly undersize and then finish reamed.

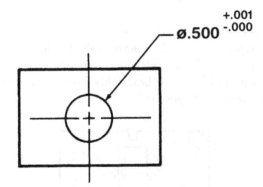

FIGURE 16.5 Reamed hole.

Boring

Boring is one of the most accurate methods for producing holes that are round, concentric, and accurately sized. *Boring* is the process of enlarging a hole with a boring tool. Boring differs from reaming in that the use of reamers is limited to the sizes of available standard reamers. Holes may be bored, however, to any desired dimension and closer tolerances, Figure 16.6.

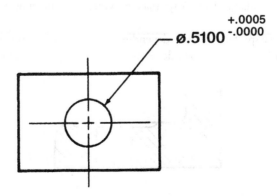

FIGURE 16.6 Bored hole.

COUNTERBORES

A counterbored hole is one that has been enlarged at one end to provide clearance for the head of a screw, bolt, or pin. The *counterbore* is usually machined to a depth that is equal to or slightly greater than the thickness of the head on the screw, bolt, or pin. This allows the head to be recessed into the surface of the workpiece, Figure 16.7.

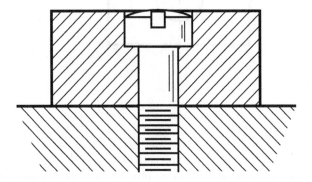

FIGURE 16.7   Counterbore.

COUNTERSINKS

A countersunk hole has a cone-shaped angle called a *countersink* cut into one end of the hole. The angle of the countersink is generally 82 degrees to match the angle found on the head of a flathead screw. Countersinks are often slightly larger than the diameter of the screw head. This allows the top of the screw head to be flush or slightly below the surface of the workpiece, Figure 16.8.

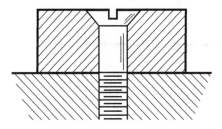

FIGURE 16.8   Countersink.

SPOTFACES

A *spotface* is similar to a counterbore, but usually not as deep, Figure 16.9. The purpose of a spotface is to provide a smooth, flat surface on an irregular surface. A spotface provides a flat bearing surface for a nut, washer, or the head of a screw or bolt.

The diameter and depth of a spotface will vary depending on the size of the nut, washer, or bolt head. Spotfaces are usually machined to a depth of 1/64 to 1/16, depending on the irregularity of the surface being machined.

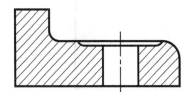

FIGURE 16.9   Spotface.

DIMENSIONING COUNTERBORES

Countersunk and counterbored holes are normally dimensioned with a leader line. Dimensions at the end of the leader line specify the minor hole diameter as well as the diameter and depth of the counterbore. The old method of calling out a counterbore is shown in Figure 16.10A. Notes and abbreviations were used to specify the diameter and depth. The new method requires the use of symbols to specify counterbore ⌴, diameter (∅), and depth ▼, Figure 16.10B.

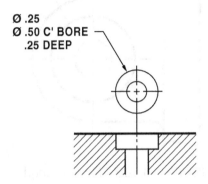

FIGURE 16.10A Old method.

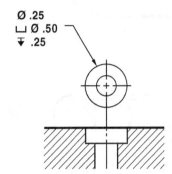

FIGURE 16.10B New method.

DIMENSIONING COUNTERSINKS

Dimensions for countersunk holes usually include the minor hole diameter, depth if not a through hole, countersink angle, and the required finished diameter of the countersink. Figures 16.11A and 16.11B show the old method and new method of calling out countersinks. Note that the old method of specifying a countersink was to use the letters CSK. The new method replaces the CSK abbreviation with the symbol ∨.

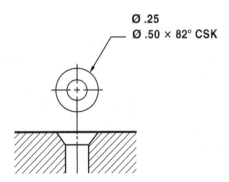

FIGURE 16.11A Old method.

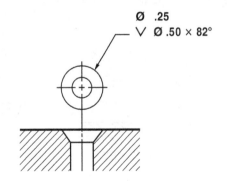

FIGURE 16.11B New method.

DIMENSIONING SPOTFACES

The diameter and depth of a spotface are usually specified on the drawing. In some cases, however, the thickness of remaining material is specified. The old method of dimensioning was to use the letters SF to specify a spotface, Figure 16.12A. Here again, symbols are used to replace notes and abbreviations, Figure 16.12B. If a dimension is not provided for the depth of the spotface or the thickness of remaining material, the spotface should be cut to the minimum depth necessary to clean up the bearing surface to the specified diameter.

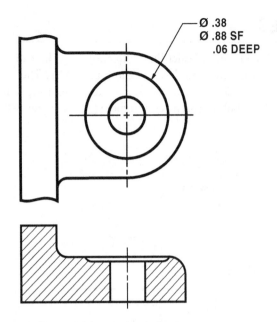

FIGURE 16.12A   Old method.

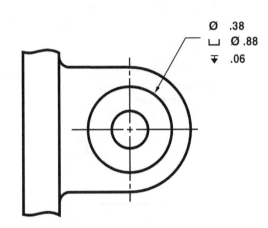

FIGURE 16.12B   New method.

KEYS, KEYWAYS, AND KEYSEATS

A *key* is a specially shaped piece of metal used to align mating parts or keep parts from rotating on a shaft. Keys are usually standard items available in various sizes.

A *keyseat* and a *keyway* are slots that are cut so the key fits into them. The slot is cut into both mating parts; a keyway is machined into the interior of a hole, while a keyseat is machined into the exterior of a shaft. Figure 16.13 shows various shaped keys and keyseats.

The dimension for a keyway or keyseat specifies the width, location, and sometimes the length. Woodruff key sizes are specified by a number.

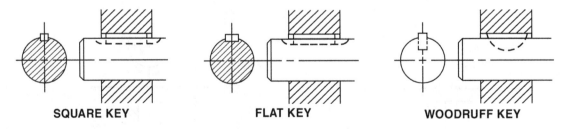

SQUARE KEY   **FLAT KEY**   **WOODRUFF KEY**

FIGURE 16.13   Keyseats and keyways.

Dimensioning Keyways and Keyseats

Keyways and keyseats are dimensioned by width, depth, location, and, if necessary, length. Common practice is to dimension the depth of the keyseat from the opposite side of the shaft and the keyway from the opposite side of the hole, Figure 16.14.

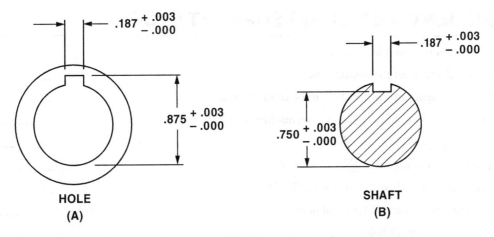

FIGURE 16.14   (A) Dimensioning keyways. (B) Dimensioning keyseats.

ASSIGNMENT D-11: SHAFT SUPPORT BLOCK

1. How many counterbored holes are there? _____

2. What is the diameter of the counterbore? _____

3. What is the diameter of the thru hole to be counterbored? _____

4. What cutting tool would likely be used to machine this hole? _____

5. What is the depth of the counterbore? _____

6. What type of line is shown at Ⓑ? _____

7. What does this line tell you about the part? _____

8. At what angle to the vertical is the oil hole? _____

9. What diameter is the oil hole? _____

10. How deep is the keyway in the right view? _____

11. What are the upper and lower limit dimensions for the width of the keyway? _____

12. What is the most likely size and shape of the standard key to be used? _____

13. What is the outside diameter of the shaft support? _____

14. The support arm for the bracket is shown in section A-A.
    What does the ☐ symbol mean? _____

15. What is the vertical distance from the centerline of the slot to
    the centerline of the shaft? _____

16. What is the horizontal distance from the surface on the back of
    the support to the centerline of the shaft? _____

17. How far from the top of the support is the centerline of the slot? _____

18. Determine distance Ⓐ. _____

19. What is the diameter of the large hole in the shaft support? _____

20. What are the limit dimensions for the large hole in the shaft support? _____

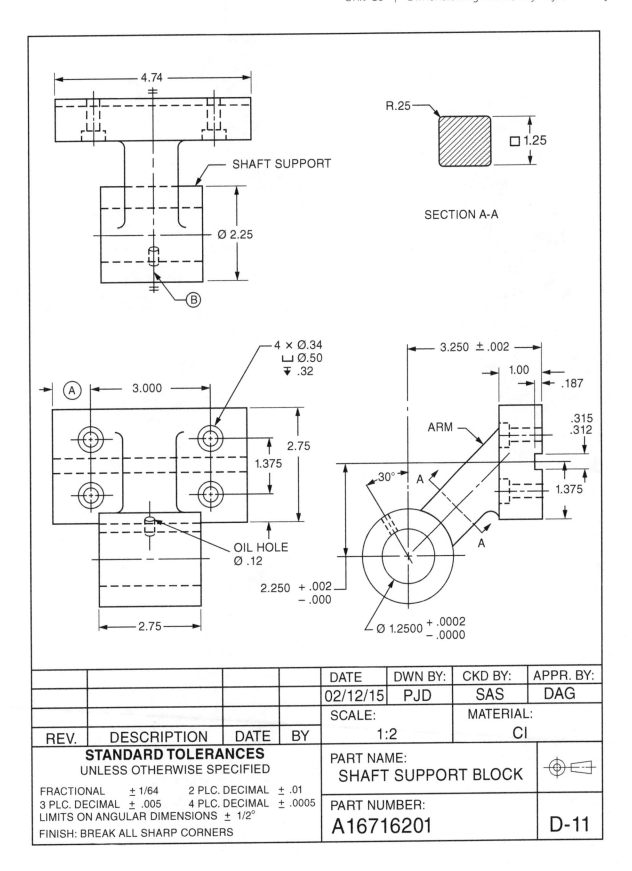

SHAFT SUPPORT

Ø 2.25

B

R.25

□ 1.25

SECTION A-A

4.74

4 × Ø.34
⌴ Ø.50
▼ .32

A   3.000

2.75

1.375

OIL HOLE
Ø .12

2.75

2.250 + .002
      − .000

3.250 ± .002

1.00

.187

ARM

.315
.312

1.375

30°   A

A

A

Ø 1.2500 + .0002
         − .0000

				DATE	DWN BY:	CKD BY:	APPR. BY:
				02/12/15	PJD	SAS	DAG
				SCALE:		MATERIAL:	
REV.	DESCRIPTION	DATE	BY	1:2		CI	

STANDARD TOLERANCES
UNLESS OTHERWISE SPECIFIED

PART NAME:
SHAFT SUPPORT BLOCK

FRACTIONAL   ± 1/64   2 PLC. DECIMAL ± .01
3 PLC. DECIMAL ± .005   4 PLC. DECIMAL ± .0005
LIMITS ON ANGULAR DIMENSIONS ± 1/2°
FINISH: BREAK ALL SHARP CORNERS

PART NUMBER:
A16716201

D-11

UNIT 17

DIMENSIONING ARCS AND RADII

SMALL ARCS

An *arc* is a portion of the circumference of a circle. Arcs are dimensioned on the view where they appear in true size and shape. The dimension given extends from the center of the arc to the circumference. This dimension is called the *radius* of the arc. The abbreviation *R* is given before the numerical dimension to indicate that it is a radius. The center of the radius is shown by a small cross. When space is limited, a radius may be shown with a leader and a dimension, as shown in Figure 17.1.

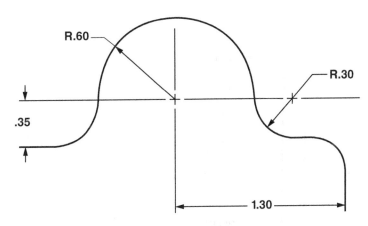

FIGURE 17.1   A small cross indicates the center of a radius.

LARGE ARCS

The radius of a large arc may fall outside the boundaries of the drawing paper. When this is the case, the radius dimension line is broken to show it is not true length, as shown in Figure 17.2. The length of a radius dimension line may also be such that it extends into another view. In that case, a broken dimension line may be used even though space on the paper is sufficient.

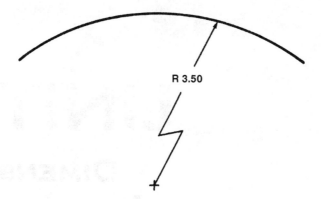

FIGURE 17.2   A broken radius dimension line shows it is not true length.

FILLETS

A *fillet* is additional metal allowed in the inner intersection of two surfaces, Figure 17.3. The fillet increases the strength of the object at a weak point where a fracture might occur. Fillets may be produced by machining, or they may be cast or forged into the part during the manufacturing process.

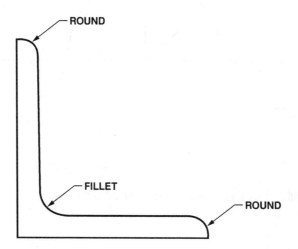

FIGURE 17.3   Fillets and rounds.

ROUNDS

A *round* is an outside radius added to a piece, Figure 17.3. A round improves the appearance of an object. It also avoids forming a sharp edge that might cause interference or chip off under a sharp blow. As with fillets, a round may be produced by machining, casting, or forging.

Note: On machined components, it is always good practice to break sharp edges for safety.

STANDARD NOTES AND TOLERANCES

Many prints provide a standardized notation for fillets and rounds rather than dimensioning each one. A typical note is shown in Figure 17.4.

> UNLESS OTHERWISE SPECIFIED ALL
> FILLETS AND ROUNDS R .12

FIGURE 17.4 Standard notes.

ASSIGNMENT D-12: SEPARATOR BRACKET

1. What scale is the drawing? _____

2. What radius is specified for the fillets and rounds on the part? _____

3. What line in the top view represents surface Ⓐ in the right side view? _____

4. What is the diameter of the spotface shown in the right side view? _____

5. What is the depth of the spotface? _____

6. What is the diameter of hole that is spotfaced? _____

7. How thick are the ribs? _____

8. Determine diameter Ⓛ. _____

9. What is distance Ⓞ? _____

10. Determine distance Ⓜ. _____

11. What is the overall height of the part at Ⓚ? _____

12. Determine distance Ⓝ. _____

13. What is distance Ⓑ? _____

14. How is distance Ⓑ determined? _____

15. What is distance Ⓒ? _____

16. What machining process would be used to produce the Ø 2.250 ± .001 hole? _____

17. What is distance Ⓔ? _____

18. What is the machining operation called when the diameter of a hole is enlarged as at Ⓗ? _____

19. What is the diameter of the hole at Ⓗ? _____

20. What is the depth of Ⓗ? _____

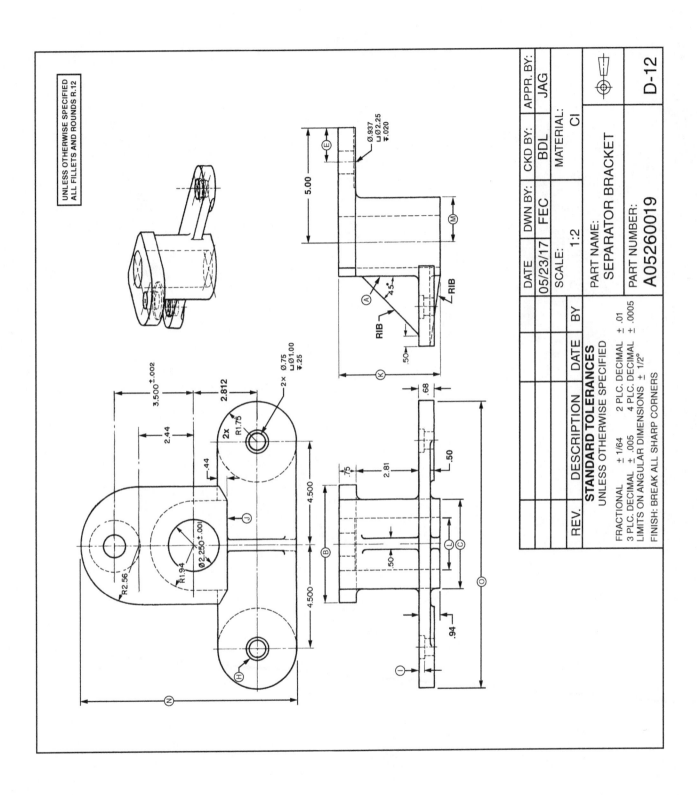

UNLESS OTHERWISE SPECIFIED
ALL FILLETS AND ROUNDS R.12

Ø.937
⌴Ø2.25
⬎.020

5.00

E

M

A

45°

RIB

RIB

.50

K

3.500 ±.002

2.812

2.44

.44

2x Ø.75
⌴Ø1.00
⬎.25

2x
R1.75

.68

.75

2.81

.50

4.500

J

B

R2.56

R1.94

Ø2.250 ±.001

.50

L

C

4.500

O

.94

H

N

I

REV.	DESCRIPTION	DATE	BY

	DATE	DWN BY:	CKD BY:	APPR. BY:
	05/23/17	FEC	BDL	JAG

SCALE:
1:2

MATERIAL:
CI

PART NAME:
SEPARATOR BRACKET

PART NUMBER:
A0526 0019

D-12

STANDARD TOLERANCES
UNLESS OTHERWISE SPECIFIED

FRACTIONAL ± 1/64   2 PLC. DECIMAL ± .01
3 PLC. DECIMAL ± .005   4 PLC. DECIMAL ± .0005
LIMITS ON ANGULAR DIMENSIONS ± 1/2°
FINISH: BREAK ALL SHARP CORNERS

UNIT 18
HOLE PATTERNS AND REVISION BLOCKS

Holes are often spaced in a circular pattern on an object. Each hole location is positioned based on the size and centerline of the circular pattern.

The circle formed by the centerline is often referred to as the ***bolt circle*** or ***bolt hole circle***. The bolt circle is dimensioned by specifying the diameter of the circle. Holes located on the hole circle may be equally spaced from each other or unequally spaced.

HOLES EQUALLY SPACED

A circle contains 360 degrees. Holes may be located around the circumference by dividing the number of holes required by the number of degrees. For example:

4 holes equally spaced on a circle = 360° ÷ 4 = 90°

In this example, the location of the holes on the circle are 90 degrees apart. The hole diameter, spacing in degrees, and the number of holes of equal size are specified by dimensions, notes, and symbols at the end of a leader line, Figure 18.1. Additionally, some drawings may have a notation or abbreviation for equal spacing such as EQL SP.

Holes of unequal size must be identified with additional dimensions and leader lines.

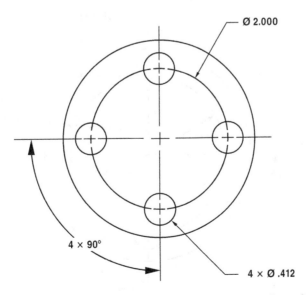

Ø 2.000

4 × 90°

4 × Ø .412

FIGURE 18.1 The notes indicate the size, quantity, and *equal* spacing of the holes.

129

HOLES UNEQUALLY SPACED

Holes unequally spaced on a circle are usually located by means of angular dimensions. The angular dimensions use a common centerline for reference to aid in proper hole location, Figure 18.2. The diameter of the bolt circle is also provided with a diagonal dimension line.

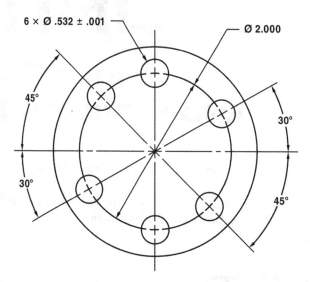

FIGURE 18.2   The notes indicate the size, quantity, and *unequal* spacing of the holes.

COORDINATE DIMENSIONS

Coordinate dimensions are taken from two perpendicular reference points. In the case of circular hole patterns, the two perpendicular centerlines are used. The hole locations are dimensioned from these points to the centerlines of each hole, Figure 18.3. The diameter of a bolt hole circle is dimensioned with a reference dimension. The numerical value of the diameter is enclosed within parentheses. Reference dimensions are given for information but should not be measured.

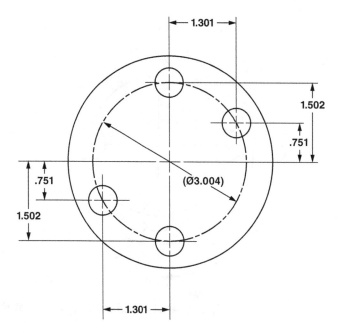

FIGURE 18.3   Coordinate dimensions.

Coordinate dimensioning is used on parts to be machined on automatic machines. They are also provided for machines equipped with digital readouts. Coordinate dimensions are used where extreme accuracy is required.

REVISIONS OR CHANGE NOTES

Completed drawings often require changes or revisions to the original design. This may be the result of product improvement, customer request, or correction of an original error. Machining processes or efforts to reduce costs will also dictate when or where a change is required. Many design changes occur after a part has already gone into production.

Revision Blocks

It is important to document all revisions or changes for future reference. When a change is made it must be recorded on the drawing and in a *revision block*. The revision block is usually located in a corner of the drawing or incorporated into the title block. The revision block information should include a revision number or letter, a description of the change that was made, the date of the change, and the name or initials of the person revising the drawing, Figure 18.4.

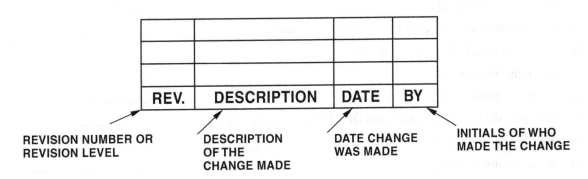

FIGURE 18.4 All changes should be documented.

Revision Symbols

Minor dimensional changes in size or location can often be made without altering original lines on a drawing. The place on a print where a change was made may be indicated by a letter or number enclosed in a circle or triangle. The change can be referenced quickly by comparing the numbers or letters on the print and those in the *revision box*, Figure 18.5. Dimension revisions may require a <u>Not To Scale</u> symbol below the revision.

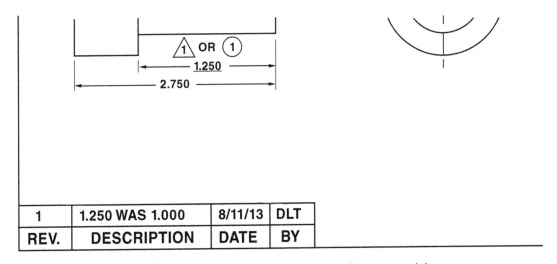

1	1.250 WAS 1.000	8/11/13	DLT
REV.	DESCRIPTION	DATE	BY

FIGURE 18.5 Circled number or numbers in a triangle indicate a revision.

ASSIGNMENT D-13: MOUNTING BRACKET

1. What is the scale of this drawing? _____

2. What revision was made at ①? _____

3. What revision was made at ②? _____

4. What is the upper limit dimension for the ⌀ 1.50 hole? _____

5. What is the lower limit dimension for the ⌀ 1.50 hole? _____

6. What does the line under the 3.88 dimension indicate? _____

7. What radius is specified for the fillets and rounds? _____

8. What is the diameter of the bolt circle for the ⌀ .44 holes? _____

9. Are the ⌀ .44 holes equally spaced? _____

10. What is the hole spacing in degrees? _____

11. What do the hidden lines in the top view show? _____

12. Determine dimension Ⓕ. _____

13. What tolerance is specified for the two place dimensions? _____

14. What is the overall thickness of the part? _____

15. Determine dimension Ⓖ. _____

16. Determine dimension Ⓙ. _____

17. What does the ( ) around dimension Ⓙ indicate? _____

18. Determine dimension Ⓗ. _____

19. Determine dimension Ⓝ. _____

20. What material is specified for the part? _____

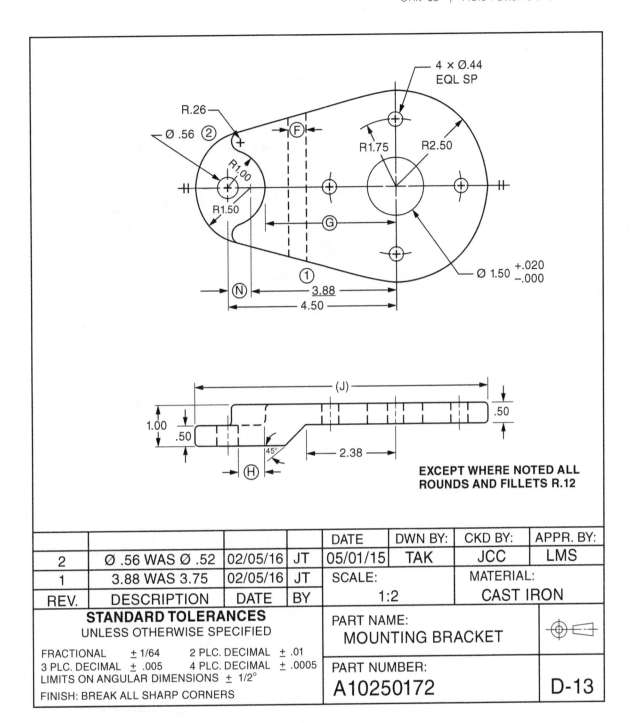

4 × Ø.44
EQL SP

R.26

Ø .56 ②

R1.00

R1.50

R1.75

R2.50

F

G

N

① 3.88

4.50

Ø 1.50 +.020 −.000

(J)

1.00

.50

.50

45°

2.38

H

EXCEPT WHERE NOTED ALL ROUNDS AND FILLETS R.12

				DATE	DWN BY:	CKD BY:	APPR. BY:
2	Ø .56 WAS Ø .52	02/05/16	JT	05/01/15	TAK	JCC	LMS
1	3.88 WAS 3.75	02/05/16	JT	SCALE:		MATERIAL:	
REV.	DESCRIPTION	DATE	BY	1:2		CAST IRON	

STANDARD TOLERANCES UNLESS OTHERWISE SPECIFIED	PART NAME: MOUNTING BRACKET	
FRACTIONAL ± 1/64 2 PLC. DECIMAL ± .01 3 PLC. DECIMAL ± .005 4 PLC. DECIMAL ± .0005 LIMITS ON ANGULAR DIMENSIONS ± 1/2° FINISH: BREAK ALL SHARP CORNERS	PART NUMBER: A10250172	D-13

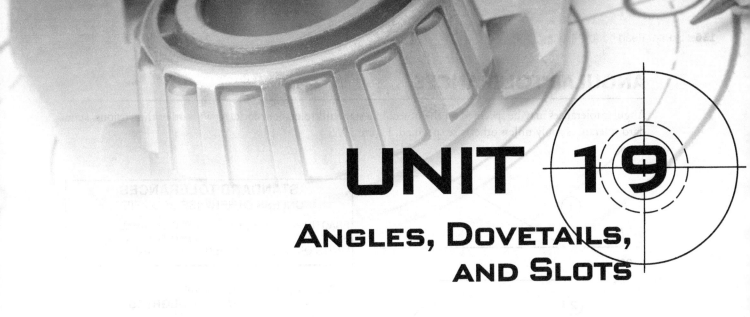

UNIT 19
ANGLES, DOVETAILS, AND SLOTS

MEASUREMENT OF ANGLES

Some objects do not have all their straight lines drawn horizontally and vertically. The design of the part may require some lines to be drawn at an angle, Figure 19.1.

The amount by which these lines diverge or draw apart is indicated by an **angle dimension**. The unit of measure of such an angle is the **degree** and is denoted by the symbol °. There are 360 degrees in a complete circle. On a drawing, 360 degrees may be written as 360°.

ANGULAR DIMENSIONS

Sizes of angles are measured in degrees. Each degree is 1/360 of a circle. The degree may be further divided into smaller units called **minutes** ('). There are 60 minutes in each degree. The minute may be further divided into smaller units called **seconds** ("). There are 60 seconds in each minute. For example, an angle measuring 10° 15′ 35″ would be a typical dimension specified in degrees, minutes, and seconds.

Angles may also be specified in decimal parts of a degree. For example, a dimension of 30° 30′ can be shown as 30.5°. Specifying angles in decimal parts of a degree is the preferred method. However, angular dimensions displayed in minutes and seconds are still found on many engineering drawings.

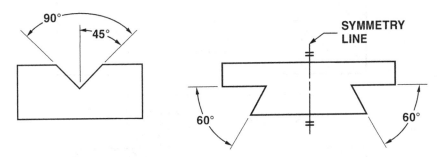

FIGURE 19.1 Machined parts with angular surfaces.

IMPLIED 90-DEGREE ANGLES

Surface features and centerlines intersecting at right angles are not specified with an angular dimension of 90 degrees on the drawing. It is generally understood that lines and surfaces shown to be at right angles will be 90 degrees unless otherwise specified. The standard tolerance for implied 90-degree angles is the same as the standard tolerance specified for other angular features on the drawing.

ANGULAR TOLERANCES

Angular tolerances may be specified on the object in a standard tolerance block. As with other dimensions, standard tolerances apply unless otherwise specified, Figure 19.2.

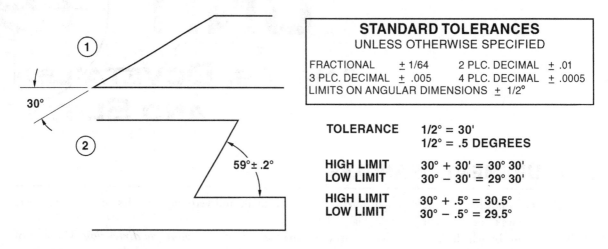

FIGURE 19.2   Angular tolerances specified on a drawing.

DOVETAILS

A *dovetail* refers to a groove or slot whose sides are cut at an angle, making an interlocking joint between two pieces so as to resist pulling apart in all directions except along the ways of the dovetail slide itself. The dovetail is commonly used in the design of slides on such machine parts as the cross slide of a lathe, the slide on the underside of a milling machine table, and for other sliding parts. The two parts of a dovetail slide are shown in cross section in Figure 19.3.

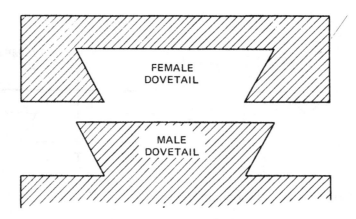

FIGURE 19.3   Two parts of dovetail slide.

The edges of a dovetail are usually broken to remove the sharp corners. On large dovetails the external and internal corners are often machined as shown in Figure 19.4 at (A) or (B).

Dovetails are usually dimensioned as shown in Figure 19.5. The dimensions limit the boundaries to which the machinist works.

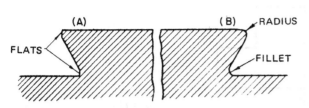

FIGURE 19.4 Types of corners used on large dovetails.

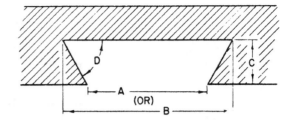

FIGURE 19.5 Dimensions of a dovetail.

MEASURING DOVETAILS

Female dovetails are measured by placing two accurate rods of known diameter against the sides and bottom of the dovetail as shown in Figure 19.6. Male dovetails may also be measured in a similar manner by placing two accurate rods of known diameter against the sides and bottom of the dovetail as shown in Figure 19.7. Acceptable limits for dimensions Q and R are calculated based on the angles of the surfaces and the size of the rods.

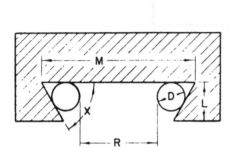

FIGURE 19.6 Measuring female dovetail.

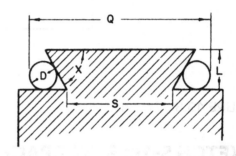

FIGURE 19.7 Measuring male dovetail.

T-SLOTS

T-slots are frequently used for clamping work securely. Drill press and milling machine tables are examples where T-slots are used. Work to be machined or work-holding devices are clamped using T-slots and T-nuts. Special cutters are used to machine T-slots.

T-NUTS

T-nuts slide along the T-slots and are typically used along with bolts to secure parts into position for machining. The T-nuts have a threaded hole so that a bolt of appropriate length can be connected. T-bolts can also be used, with their heads sliding in the T-slot, with the bolt extending, to reach clamps that are used to secure parts.

As with dovetails, T-slots and T-nuts are symmetrical and are displayed with a symmetry line, Figure 19.8.

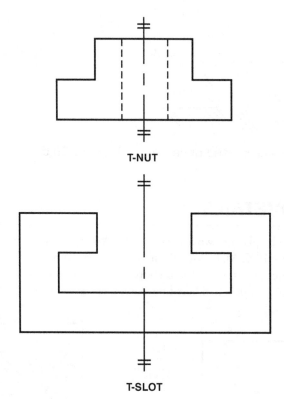

FIGURE 19.8 T-nut and T-slot.

SKETCH S-10: REST BRACKET

1. Lay out front, right-side, and top views.

2. Start the sketch .50 inch from the left-hand margin and about .50 inch from the bottom. Make the views 1.00 inch apart.

3. Dimension the completed drawing.

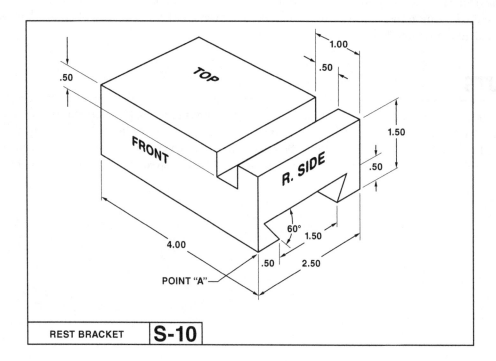

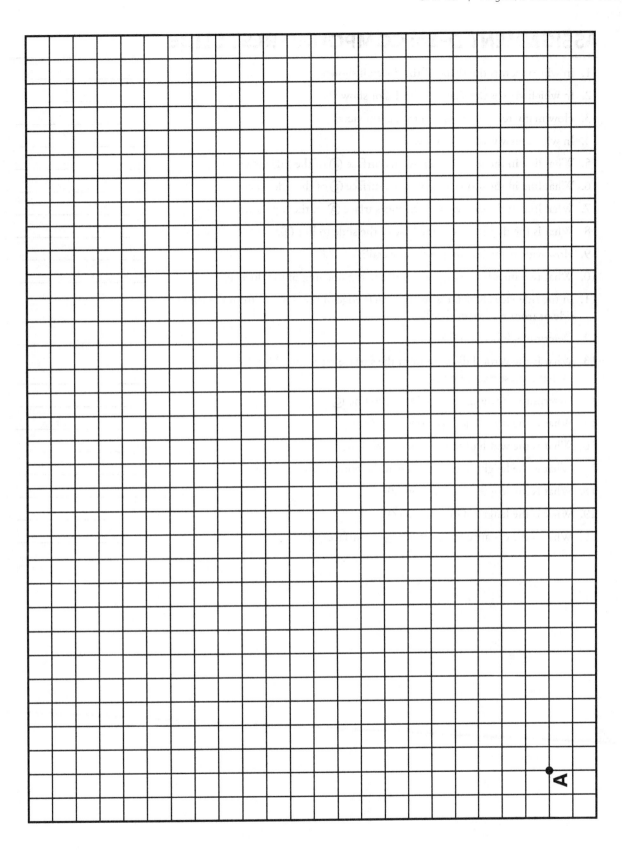

ASSIGNMENT D-14: COMPOUND REST SLIDE

1. In which view is the shape of the dovetail shown? _____

2. In which view is the shape of the T-slot shown? _____

3. How many rounds are shown in the top view? _____

4. In which view is the fillet shown? _____

5. What line in the top view represents surface Ⓡ of the side view? _____

6. What line in the top view represents surface Ⓛ of the side view? _____

7. What line in the side view represents surface Ⓐ of the top view? _____

8. What is the distance from the base of the slide to line Ⓙ? _____

9. How wide is the opening in the dovetail? _____

10. What two lines in the top view indicate the opening of the dovetail? _____

11. In the side view, how far is the lower left edge of the dovetail from the left side or front surface of the slide? _____

12. Determine dimension Ⓨ. _____

13. What is the vertical distance from the surface represented by the line Ⓠ to that represented by line Ⓣ. _____

14. Determine the distance from line Ⓕ to line Ⓖ. _____

15. What is the overall depth of the T-slot? _____

16. What is the width of the bottom of the T-slot? _____

17. What is the height of the opening at the bottom of the T-slot? _____

18. What is the length of dimension Ⓥ? _____

19. What is the length of dimension Ⓧ? _____

20. What is the distance from surface Ⓝ to surface Ⓢ? _____

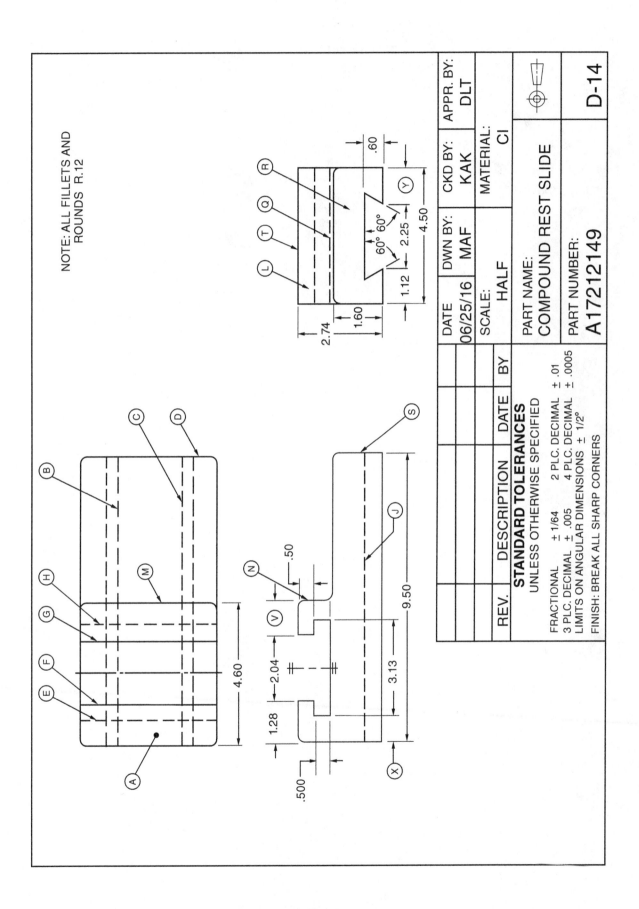

NOTE: ALL FILLETS AND
ROUNDS R.12

	DATE 06/25/16	DWN BY: MAF	CKD BY: KAK	APPR. BY: DLT
	SCALE: HALF		MATERIAL: CI	
	PART NAME: COMPOUND REST SLIDE			
	PART NUMBER: A17212149			D-14

REV.	DESCRIPTION	DATE	BY

STANDARD TOLERANCES
UNLESS OTHERWISE SPECIFIED

FRACTIONAL   ± 1/64   2 PLC. DECIMAL   ± .01
3 PLC. DECIMAL ± .005   4 PLC. DECIMAL   ± .0005
LIMITS ON ANGULAR DIMENSIONS ± 1/2°

FINISH: BREAK ALL SHARP CORNERS

UNIT 20

METRIC DIMENSIONS AND TOLERANCES

INTERNATIONAL SYSTEM OF METRIC UNITS (SI)

In 1954, the Conférence Générale des Poids et Mesures (CGPM), which is responsible for all international metric decisions, adopted the Système International d'Unités. The abbreviation for this system is simply **SI**. The SI metric system establishes the meter as the basic unit of length. Additional length measures are formed by multiplying or dividing the meter by powers of 10, Figure 20.1.

The standard inch measurements on metric drawings are replaced by millimeter dimensions, Figure 20.2. A millimeter is 1/1000 of a meter. Angular dimensions and tolerances remain unchanged in the metric system because degrees, minutes, and seconds are common to both systems of measurement.

1 meter	$= \dfrac{1}{1000}$ kilometers
1 meter	$=$ 10 decimeters
1 meter	$=$ 100 centimeters
1 meter	$=$ 1000 millimeters

FIGURE 20.1 Length measurements based on the meter.

IN		MM
.0001	=	.00254
.001	=	.02540
.010	=	.25400
.100	=	2.54000
1.000	=	25.40000
10.000	=	254.00000

FIGURE 20.2 Inches to millimeters.

U.S. CONVERSION TO SI METRICS (SI)

The metric system of measurement is certainly not new to the world. It was first established in France in the late 1700s and has since become the standard of measurement in most countries. However, the traditional system of measurement most familiar in the United States is the U.S. Customary system of measurement.

However, in the mid 1970's increased international trade and worldwide use of metrics caused a reevaluation of the U.S. Customary system. As a result, it was determined that to be competitive in foreign markets and assure interchangeability of parts, the United States should adopt the metric system. Therefore, in 1975,

President Gerald Ford signed the U.S. Metric Conversion Act into law. The law, however, specified that implementation was to be voluntary rather than required. Since that time, for a variety of reasons such as cost or a general resistance to change, the United States has been slow to fully replace the U.S. Customary system with the SI Metric system.

DIMENSIONING METRIC DRAWINGS

The millimeter is the standard linear metric unit used on engineering drawings. Dimensions may be shown as a whole number or as a number followed by a decimal point with numerals to the right of the decimal specifying the amount of precision required, Figure, 20.3A. Millimeters having a value of less than 1 are shown with a decimal point and a zero to the left of the dimension, Figure 20.3B. Dimensions greater than 1 millimeter do not require a zero to the left of the number.

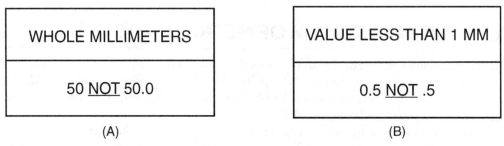

WHOLE MILLIMETERS	VALUE LESS THAN 1 MM
50 <u>NOT</u> 50.0	0.5 <u>NOT</u> .5
(A)	(B)

FIGURE 20.3   Dimensioning metric drawings.

Metric drawings should be labeled as metric in the title block area or on the drawing itself in bold letters, Figure 20.4. Additionally, a note such as "**UNLESS OTHERWISE SPECIFIED ALL DIMENSIONS ARE IN MILLIMETERS**" may be used. Angular dimensions are shown in degrees or decimal parts of a degree.

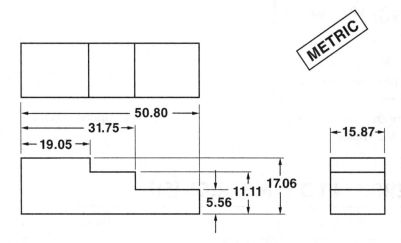

FIGURE 20.4   Drawing dimensioned in metric units of measure only.

METRIC TOLERANCES

Information in the standard tolerance block found on most engineering drawings will display the standard metric tolerance applied. Tolerances applied to size or location dimensions on the drawing usually specify closer control of the dimension. Metric tolerances applied to dimensions may be shown as bilateral or unilateral. A unilateral tolerance will show the +0 or -0 simply as 0. Bilateral tolerances are shown in the same way that decimal tolerances are specified, Figure 20.5.

UNILATERAL TOLERANCE	UNILATERAL TOLERANCE	BILATERAL TOLERANCE	BILATERAL TOLERANCE
50 $^{\ \ \ 0}_{-\ 0.2}$	50 $^{+\ 0.5}_{\ \ \ 0}$	102 ± 0.2	102 $^{+\ 0.5}_{-\ 0.2}$

FIGURE 20.5 Metric tolerances.

DUAL DIMENSIONS

Although most engineering drawings are either decimal or metric, some may specify dimensions using both systems. This practice is called dual dimensioning. ***Dual dimensions*** provide a reference when converting from one system to the other. When dual dimensions are used to specify a dimension, brackets should be shown around the secondary unit value, Figure 20.6.

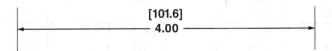

FIGURE 20.6 Dual dimensions.

DECIMAL AND MILLIMETER CONVERSION

Decimal dimensions can be converted to millimeters by multiplying the decimal value by 25.4.

> **EXAMPLE:** .750 IN X 25.4 MM = 19.050 MM

Dimensions shown in millimeters can be converted to decimals by multiplying the decimal value by .03937.

> **EXAMPLE:** 19.050 MM X .03937 IN = .750 IN

Note: A Metric/English (Decimal) Conversion Chart is shown on the following page.

MILLIMETER-DECIMAL CONVERSION TABLE

Millimeter	Decimal	Millimeter	Decimal	Millimeter	Decimal	Millimeter	Decimal	Millimeter	Decimal
0.1	.0039	5.159	.2031	10.2	.4016	15.3	.6024	20.3	.7992
0.2	.0079	5.2	.2047	10.3	.4055	15.4	.6083	20.4	.8031
0.3	.0118	5.3	.2087	10.319	.4063	15.478	.6094	20.5	.8071
0.397	.0156	5.4	.2128	10.4	.4094	15.5	.6102	20.6	.8110
0.4	.0157	5.5	.2165	10.5	.4134	15.6	.6142	20.638	.8125
0.5	.0197	5.556	.2188	10.6	.4173	15.7	.6181	20.7	.8150
0.6	.0236	5.6	.2205	10.7	.4213	15.8	.6220	20.8	.8189
0.7	.0276	5.7	.2244	10.716	.4219	15.875	.6250	20.9	.8228
0.794	.0313	5.8	.2283	10.8	.4252	15.9	.6260	21.0	.8268
0.8	.0315	5.9	.2323	10.9	.4291	16.0	.6299	21.034	.8281
0.9	.0354	5.953	.2344	11.0	.4331	16.1	.6339	21.1	.8307
1.0	.0394	6.0	.2362	11.1	.4370	16.2	.6378	21.2	.8346
1.1	.0433	6.1	.2402	11.113	.4375	16.272	.6408	21.3	.8386
1.191	.0469	6.2	.2441	11.2	.4409	16.3	.6417	21.4	.8425
1.2	.0472	6.3	.2480	11.3	.4449	16.4	.6457	21.431	.8438
1.3	.0512	6.350	.2500	11.4	.4488	16.5	.6496	21.5	.8465
1.4	.0551	6.4	.2520	11.5	.4528	16.6	.6535	21.6	.8504
1.5	.0591	6.5	.2559	11.509	.4531	16.669	.6563	21.7	.8543
1.588	.0625	6.6	.2598	11.6	.4567	16.7	.6575	21.8	.8583
1.6	.0630	6.7	.2638	11.7	.4606	16.8	.6614	21.828	.8594
1.7	.0669	6.747	.2656	11.8	.4646	16.9	.6654	21.9	.8822
1.8	.0709	6.8	.2677	11.9	.4685	17.0	.6893	22.0	.8661
1.9	0.748	6.9	.2717	11.906	.4688	17.055	.6719	22.1	.8701
1.984	.0781	7.0	.2756	12.0	.4724	17.1	.6732	22.2	.8740
2.0	.0787	7.1	.2795	12.1	.4764	17.2	.6772	22.225	.8750
2.1	.0827	7.144	.2813	12.2	.4803	17.3	.6811	22.3	.8780
2.2	.0866	7.2	.2835	12.3	.4843	17.4	.6850	22.4	.8819
2.3	.0906	7.3	.2874	12.303	.4844	17.463	.6875	22.5	.8858
2.381	.0938	7.4	.2913	12.4	.4882	17.5	.6890	22.6	.8898
2.4	.0945	7.5	.2953	12.5	.4921	17.6	.6929	22.622	.8906
2.5	.0984	7.541	.2989	12.6	.4961	17.7	.6968	22.7	.8937
2.6	.1024	7.6	.2992	12.7	.5000	17.8	.7008	22.8	.8976
2.7	.1063	7.7	.3031	12.8	.5039	17.859	.7031	22.9	.9016
2.778	.1094	7.8	.3071	12.9	.5079	17.9	.7047	23.0	.9055
2.8	.1102	7.9	.3110	13.0	.5118	18.0	.7087	23.019	.9063
2.9	.1142	7.938	.3125	13.097	.5156	18.1	.7126	23.1	.9094
3.0	.1181	8.0	.3150	13.1	.5157	18.2	.7165	23.2	.9134
3.1	.1220	8.1	.3189	13.2	.5197	18.256	.7188	23.3	.9173
3.175	.1250	8.2	.3228	13.3	.5236	18.3	.7205	23.4	.9213
3.2	.1260	8.3	.3268	13.4	.5276	18.4	.7244	23.416	.9219
3.3	.1299	8.334	.3281	13.494	.5313	18.5	.7283	23.5	.9252
3.4	.1339	8.4	.3307	13.5	.5315	18.6	.7323	23.6	.9291
3.5	.1378	8.5	.3346	13.6	.5354	18.653	.7344	23.7	.9331
3.572	.1406	8.6	.3386	13.7	.5394	18.7	.7362	23.8	.9370
3.6	.1417	8.7	.3425	13.8	.6433	18.8	.7402	23.813	.9375
3.7	.1457	8.731	.3438	13.891	.5469	18.9	.7441	23.9	.9409
3.8	.1496	8.8	.3465	13.9	.5472	19.0	.7480	24.0	.9449
3.9	.1535	8.9	.3504	14.0	.5512	19.050	.7500	24.1	.9488
3.969	.1563	9.0	.3543	14.1	.5551	19.1	.7520	24.2	.9528
4.0	.1575	9.1	.3583	14.2	.5591	19.2	.7559	24.209	.9531

4.1	.1614	9.128	.3594	14.288	.5625	19.3	.7598	24.3	.9567
4.2	.1654	9.2	.3622	14.3	.5630	19.4	.7638	24.4	.9606
4.3	.1693	9.3	.3661	14.4	.5669	19.447	.7656	24.5	.9646
4.366	.1719	9.4	.3701	14.5	.5709	19.5	.7677	24.6	.9685
4.4	.1732	9.5	.3740	14.6	.5748	19.6	.7717	24.608	.9688
4.5	.1772	9.525	.3750	14.684	.5781	19.7	.7756	24.7	.9724
4.6	.1811	9.6	.3780	14.7	.5787	19.8	.7795	24.8	.9764
4.7	.1850	9.7	.3819	14.8	.5827	19.844	.7813	24.9	.9803
4.763	.1875	9.8	.3858	14.9	.5666	19.9	.7835	25.0	.9843
4.8	.1890	9.9	.3898	15.0	.5906	20.0	.7874	25.003	.9844
4.9	.1929	9.922	.3906	15.081	.5938	20.1	.7913	25.1	.9882
5.0	.1969	10.0	.3937	15.1	.5945	20.2	.7953	25.2	.9921
5.1	.2008	10.1	.3976	15.2	.5984	20.241	.7969	25.3	.9961
								25.400	1.0000

FIGURE 20.7 Decimal and millimeter conversion.

ASSIGNMENT D-15: ADAPTER

1. What is the name of the system of measurement that uses inch units? _____

2. What is the official name given to the metric system of measurement? _____

3. What is the diameter of the bolt circle on the Adapter? _____

4. What is the hole spacing in degrees? _____

5. How many 8-mm holes are required? _____

6. Use the attached metric conversion chart to convert the metric dimensions on the Adapter rounded to two places. _____

    8 mm = _____

    15 mm = _____

    16 mm = _____

    20 mm = _____

    32 mm = _____

    40 mm = _____

    50 mm = _____

    60 mm = _____

    80 mm = _____

7. Using the attached metric conversion chart, convert 3 inches to millimeters. _____

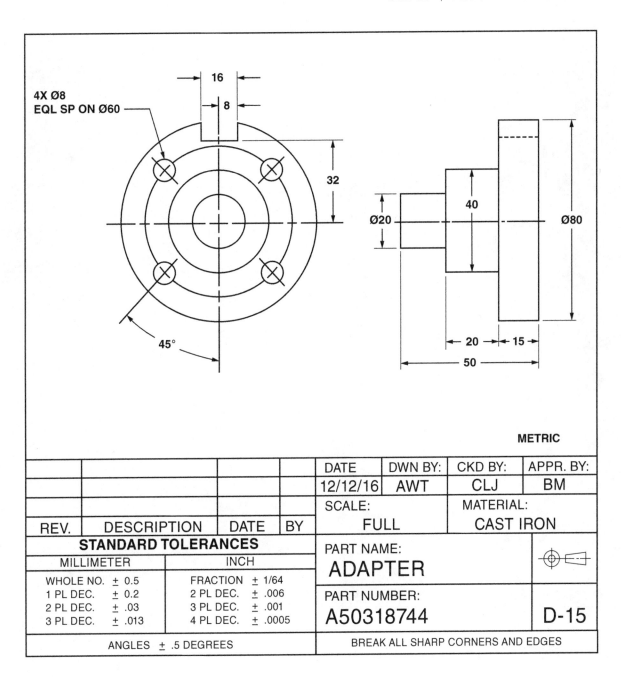

4X Ø8
EQL SP ON Ø60

16
8
32
45°

Ø20
40
Ø80

20
15
50

METRIC

DATE	DWN BY:	CKD BY:	APPR. BY:
12/12/16	AWT	CLJ	BM

SCALE:	MATERIAL:
FULL	CAST IRON

REV.	DESCRIPTION	DATE	BY

STANDARD TOLERANCES	
MILLIMETER	INCH
WHOLE NO. ± 0.5	FRACTION ± 1/64
1 PL DEC. ± 0.2	2 PL DEC. ± .006
2 PL DEC. ± .03	3 PL DEC. ± .001
3 PL DEC. ± .013	4 PL DEC. ± .0005
ANGLES ± .5 DEGREES	

PART NAME:
ADAPTER

PART NUMBER:
A50318744

D-15

BREAK ALL SHARP CORNERS AND EDGES

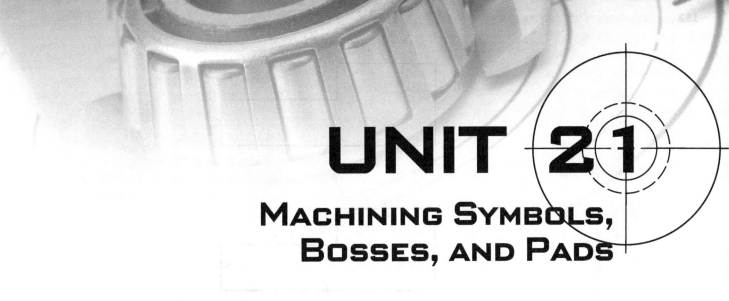

UNIT 21

MACHINING SYMBOLS, BOSSES, AND PADS

MACHINING SYMBOLS

Objects produced by a casting or forging process often have one or more rough surfaces that must be finished by machining. To allow for this, the drafter will specify the addition of extra material to the areas where a machine finish is required. Symbols called *machining symbols* are added to the drawing to show where material is to be removed. A machining symbol is "V" shaped with a long leg on the right side and a short leg on the left side. A short horizontal line extends between the two legs forming a small triangle, Figure 21.1.

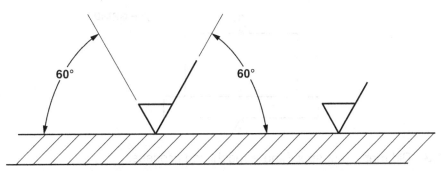

SURFACE TO BE FINISHED BY MACHINING. EXTRA MATERIAL MUST BE PROVIDED FOR MACHINING.

FIGURE 21.1 Machining symbols.

This standardized machining symbol replaces previous machining symbols, often called *finish marks*, that may still appear on drawings produced prior to the new standards as shown in Figure 21.2. Older symbols may also include a letter or word specifying the machining process to be used. For example, Figure 21.3 shows the addition of a letter "G" specifying that a grind process is required.

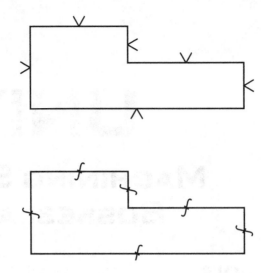

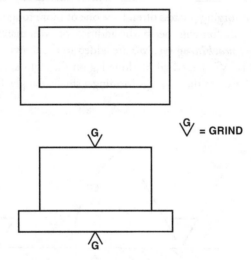

FIGURE 21.2   Old style surface finish marks.

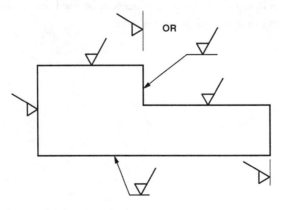

FIGURE 21.3   Old method of specifying a machining process.

PLACEMENT OF MACHINING SYMBOLS

Machining symbols are placed with the point of the "V" touching the finished surface, Figure 21.4. They may also appear on an extension line or leader line when this is convenient. As with dimensions, finish marks should only appear once on a blueprint and should not be duplicated from one view to another. Additionally, the machining symbols should be oriented so that they are read from the right-hand side or the bottom of the view.

FIGURE 21.4   New style machining symbols.

MATERIAL REMOVAL SPECIFICATIONS

Additional information specifying the amount of material to be removed by machining may be added to a machining symbol. When this is the case, a numerical value is placed to the left of the machining symbol. This value will be expressed in decimals, or in the case of a metric drawing it will be in millimeters. Figure 21.5 shows an example where .03 inch has been specified for machining.

.03   INDICATES THE AMOUNT OF MATERIAL TO BE REMOVED BY MACHINING.

FIGURE 21.5

PROHIBITING MATERIAL REMOVAL

When it is necessary to show that removal of material from a specific area of a part is prohibited, a different variation of a machining symbol is used. This symbol, when placed on the object, specifies that material may not be removed regardless of any other machining process performed. A *material removal prohibited* symbol has a small circle placed in the "V" and the horizontal line removed, Figure 21.6.

INDICATES THAT MATERIAL REMOVAL IS NOT ALLOWED. USUALLY FOUND ON CAST OR FORGED SURFACES.

FIGURE 21.6   Material removal prohibited.

BOSSES AND PADS

Bosses and pads are raised surfaces that are often found on castings to add strength, provide a machining surface, or both. A *boss* is cylindrical in shape of relatively small size and a *pad* is a noncircular projection of any shape except round, Figures 21.7A and 21.7B.

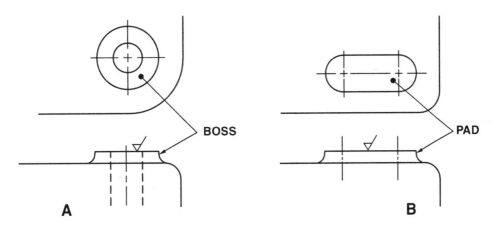

FIGURE 21.7   Bosses and pads.

A machined boss or pad provides a smooth surface for mating parts or a flat area for the head of a bolt or screw used to connect one part to another.

ASSIGNMENT D-16: RACK COLUMN BRACKET

1. How many holes are shown on the drawing?   _____

2. What is the height of the base, not including the height of the bosses?   _____

3. What is the height of the bosses above the top of the base?   _____

4. What is the radius on the corners of the base?   _____

5. What is the diameter of the bosses on the base?   _____

6. Determine distance Ⓐ.   _____

7. Determine distance Ⓑ.   _____

8. How far from the horizontal centerline in the top view are the bosses?   _____

9. Determine the distance from the pad to the centerline of the
   upright support.   _____

10. What does the ▽ symbol specify?   _____

11. How far is the centerline of the 1.002/1.000 diameter hole from the centerline
    of the ∅ .688 hole?   _____

12. What size is the hole that is parallel to the base?   _____

13. What is the distance from the bottom of the base to the center
    of the ∅ .688?   _____

14. What is the upper limit dimension of the hole in the upright?   _____

15. What is the lower limit dimension of the hole in the upright?   _____

16. What is the outside diameter of the upright?   _____

17. What does the **.02** ▽ symbol specify?   _____

18. How wide is the pad?   _____

19. What change to the height of the boss was made at ①?   _____

20. How many ∅ .41 holes are required?   _____

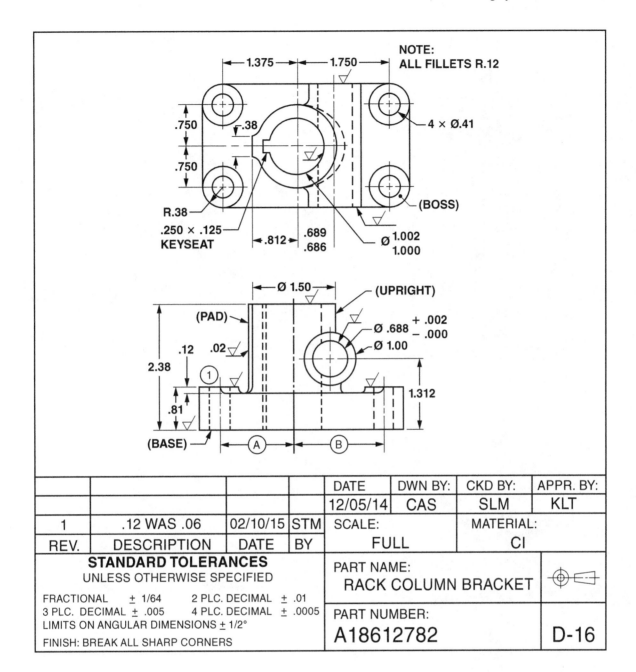

NOTE:
ALL FILLETS R.12

1.375   1.750

4 × Ø.41

.750

.38

.750

(BOSS)

R.38

.250 × .125
KEYSEAT

.812

.689
.686

Ø 1.002
1.000

Ø 1.50

(UPRIGHT)

(PAD)

Ø .688 + .002 − .000

Ø 1.00

2.38

.12   .02

1

.81

1.312

(BASE)

A   B

				DATE	DWN BY:	CKD BY:	APPR. BY:
				12/05/14	CAS	SLM	KLT
1	.12 WAS .06	02/10/15	STM	SCALE:		MATERIAL:	
REV.	DESCRIPTION	DATE	BY	FULL		CI	

STANDARD TOLERANCES
UNLESS OTHERWISE SPECIFIED

FRACTIONAL   ± 1/64   2 PLC. DECIMAL ± .01
3 PLC. DECIMAL ± .005   4 PLC. DECIMAL ± .0005
LIMITS ON ANGULAR DIMENSIONS ± 1/2°
FINISH: BREAK ALL SHARP CORNERS

PART NAME:
RACK COLUMN BRACKET

PART NUMBER:
A18612782

D-16

UNIT 22
SURFACE TEXTURE

Surface texture refers to the degree of roughness allowed on a machined surface. Modern technology demands close tolerances, high speeds, and increased resistance to friction and wear. How the part will be used is an important consideration when determining the required surface texture. For some components, a rough machined surface may be good enough. Other components may require highly polished surfaces with few surface irregularities. Surface control should only be applied to drawings where it is essential to the part. The unnecessary control of texture may lead to increased production costs. If the fit or function will be affected, then surface texture will need additional definition.

Drafters and designers must determine the appropriate surface texture required to ensure maximum performance at the lowest cost. When control of surface texture quality will need to be maintained, a surface texture symbol is used. These symbols describe the allowable roughness, waviness, height of surface irregularities, and other characteristics required on the finished part, Figure 22.1. This unit describes the various types of surface texture symbols, surface texture terminology, and where surface texture control requirements are placed on the symbols.

SURFACE TEXTURE TERMINOLOGY

Lay. Refers to the predominant direction of surface roughness caused by the machining process. The chart in Figure 22.1 provides a definition of each lay symbol.

Roughness. High and low points on a surface. These are often caused by the machining process used to generate the surface.

Roughness Height (Ra). An arithmetical average height as measured from the mean line of the roughness profile. The mean line is a point halfway between a peak and valley. Roughness height is the amount of deviation from that mean line. This number may be expressed in microinches, micrometers, or ISO (International Standards Organization) grade numbers ranging from N1 to N12.

Roughness Width. The distance between a point on a ridge to an equal point on the next ridge.

Roughness Width Cutoff. The distance of surface roughness to be included in calculating average roughness height.

Waviness. The larger undulations of a surface that lie below the surface roughness marks. Roughness and lay characteristics are imposed on top of surface waviness.

Waviness Width. A distance measured in the same way as roughness width.

Waviness Height.   Distance between the mean roughness line measured at the top and bottom of the wave.

Flaws.   Flaws are unintended surface irregularities that may appear infrequently on a finished surface. Examples include scratches, pinholes, cracks, checks, and so on.

SURFACE TEXTURE CHARACTERISTICS

The enlarged view of the machined surface in Figure 22.1 displays the typical characteristics defined in the terminology section. Standard surface texture symbols are used to ensure conformity among manufacturers.

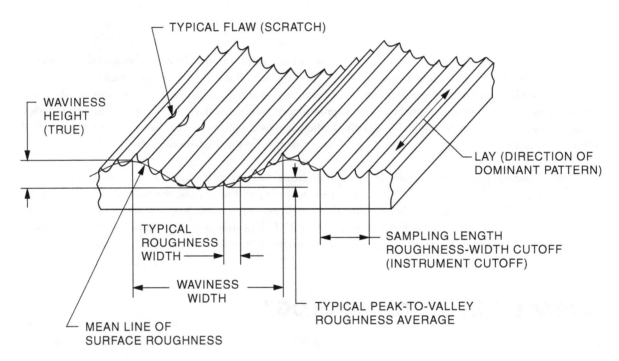

FIGURE 22.1   Surface texture characteristics.

LAY SYMBOLS

Lay symbols are used to represent the direction of the tool marks on a specified surface. Lay is specified for both function and appearance. Wear, friction, and lubricating qualities may be affected by lay. Figure 22.2 shows various lay symbols and how the tool marks might appear on the work surface.

SYMBOL	DESCRIPTION	EXAMPLE
=	LAY PARALLEL TO THE LINE REPRESENTING THE SURFACE TO WHICH THE SYMBOL IS APPLIED	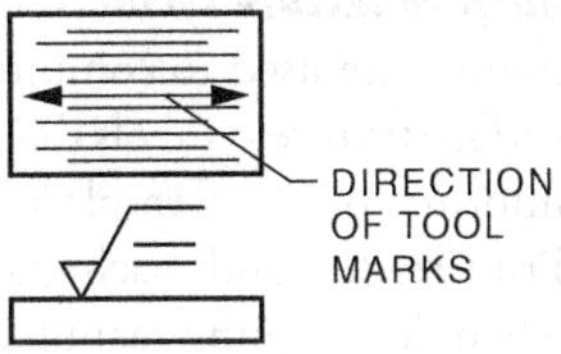 DIRECTION OF TOOL MARKS
⊥	LAY PERPENDICULAR TO THE LINE REPRESENTING THE SURFACE TO WHICH THE SYMBOL IS APPLIED	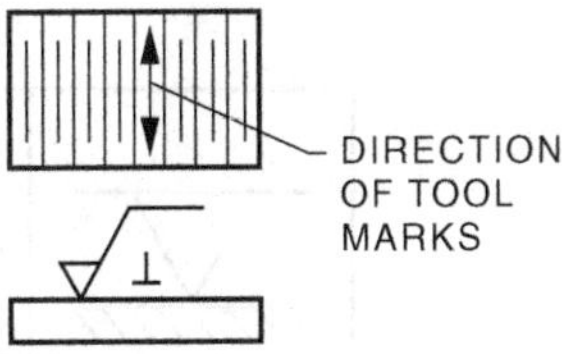 DIRECTION OF TOOL MARKS
X	LAY ANGULAR IN BOTH DIRECTIONS TO THE LINE REPRESENTING THE SURFACE TO WHICH THE SYMBOL IS APPLIED	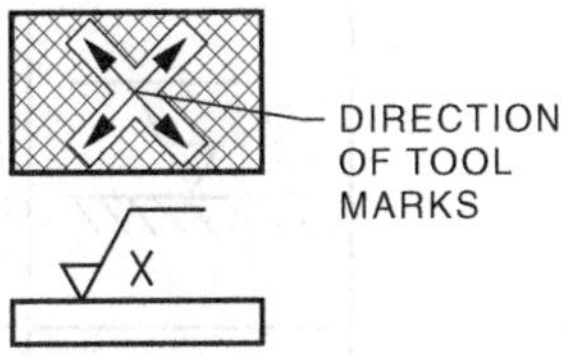 DIRECTION OF TOOL MARKS
M	LAY MULTIDIRECTIONAL	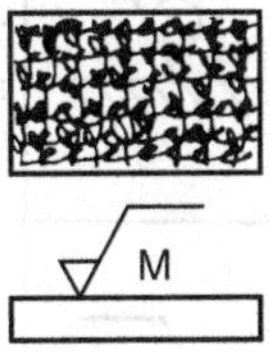
C	LAY APPROXIMATELY CIRCULAR RELATIVE TO THE CENTER OF THE SURFACE TO WHICH THE SYMBOL IS APPLIED	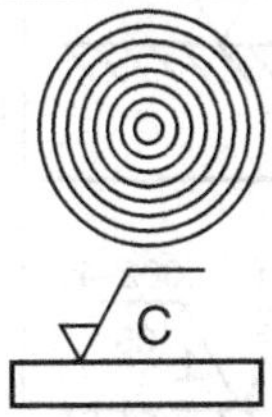
R	LAY APPROXIMATELY RADIAL RELATIVE TO THE CENTER OF THE SURFACE TO WHICH THE SYMBOL IS APPLIED	
P	LAY NONDIRECTIONAL, PITTED, OR PROTUBERANT	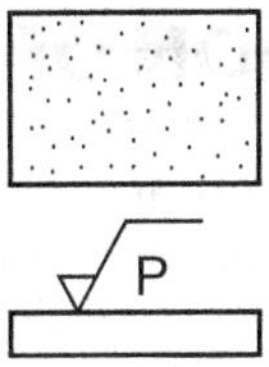

FIGURE 22.2 Lay symbols.

SURFACE TEXTURE SYMBOLS

Surface texture symbols may appear as a machining symbol (described in Unit 21). However, variations of the symbol are used to communicate and clarify surface texture control requirements. Figure 22.3 shows the basic surface texture symbols that may be found on engineering drawings and provides a description of each. Additional information is often placed around a surface texture symbol to specify greater control of finish requirements. Print readers and machinists need to be able to interpret the information provided by a surface texture symbol when determining manufacturing or machining process.

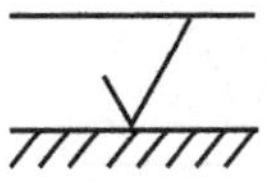	**SURFACE MAY BE PRODUCED BY ANY METHOD** This symbol indicates that the surface texture may be produced by machining or other manufacturing processes such as bending, forming or plating. A roughness average number may be added as a control feature.
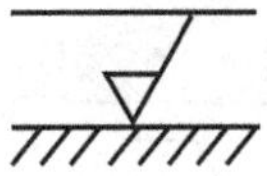	**MATERIAL REMOVAL REQUIRED** This symbol indicates that the surface is to be produced by machining. A machining allowance number may appear to the left of the symbol. This symbol does not specify surface finish.
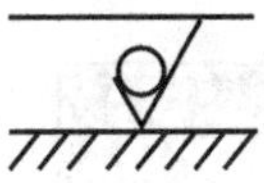	**MATERIAL REMOVAL PROHIBITED** This symbol indicates that the surface will be controlled by the manufacturing process and material removal is not allowed. The surface texture created by a casting process is a good example.
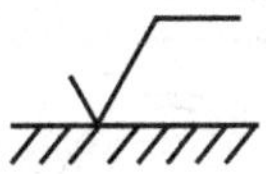	**COMPLETE SYMBOL** A complete surface texture symbol appears with a horizontal line at the top. Numbers and lay symbols are placed around the symbol in specific locations to communicate surface finish specifications.
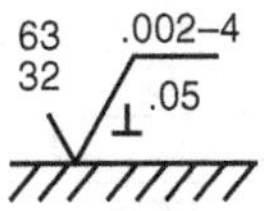 63 32 .002–4 .05	**SYMBOL WITH SURFACE TEXTURE CONTROL** A complete surface texture symbol appears with a horizontal line at the top. Numbers and lay symbols are placed around the symbol in specific locations to communicate surface finish specifications.

FIGURE 22.3 Surface texture symbols.

LOCATION OF RATINGS AND SYMBOLS

Ratings, symbols, and allowances are placed around the surface texture symbol in specific locations. Figure 22.4 illustrates the proper location for each.

Surface texture symbols, like dimensions and machining symbols, should only appear once on a drawing. They should not be duplicated on adjacent views of the same surface. When placed on a drawing, the symbol always is shown in an upright position, Figure 22.5.

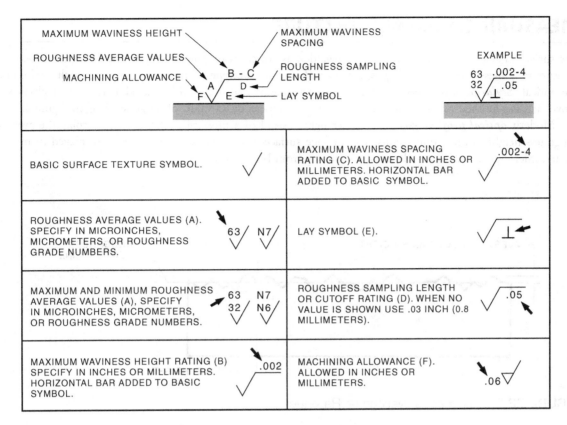

FIGURE 22.4 Location of surface texture information.

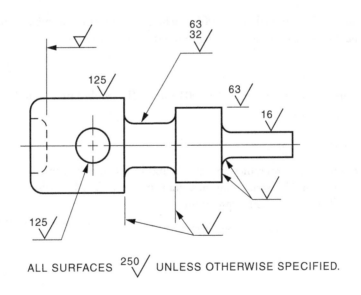

ALL SURFACES $^{250}\!\!\diagup$ UNLESS OTHERWISE SPECIFIED.

NOTE: VALUES SHOWN ARE IN MICROINCHES.

FIGURE 22.5 Application of surface texture symbols.

MEASURING SURFACE TEXTURE

The surface characteristic that is most often regarded as critical is roughness height. The instruments used to measure roughness height are called *profilometers*. Two common types of profilometers used are the stylus and the optical. *Stylus profilometers* are mechanical instruments that use a probe, called a stylus, that physically moves across a finished surface to record the roughness average, or Ra value, of roughness height, Figure 22.6.

Modern *optical profilometers* use a laser light instead of a physical probe and can display the average roughness height in three dimensions. The average surface roughness height recorded is displayed in microinches, micrometers, or ISO grade numbers ranging from N1 to N12.

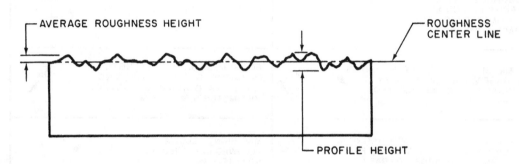

FIGURE 22.6   Roughness average or Ra value.

Microinch

A *microinch* is one millionth of an inch, or .000001 inches. The higher the number of microinches, the rougher the surface. The symbol used to indicate a reading in microinches is μin.

Micrometer

A *micrometer* is one millionth of a meter, or .000001 meter. The symbol used to indicate a reading in micrometers is μm.

Roughness Average Values

Figure 22.7 shows recommended roughness average ratings in microinches, micrometers, and in ISO (International Standards Organization) "N" series roughness grade numbers. The ISO system may be found on drawings exchanged internationally to avoid misinterpretation.

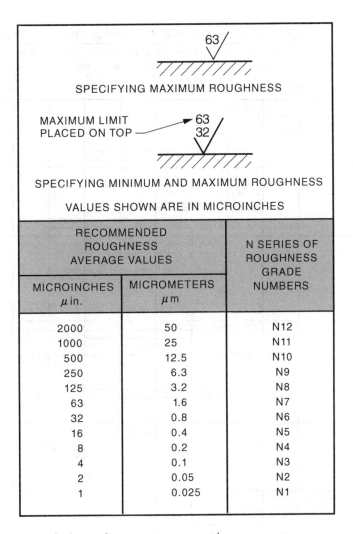

SPECIFYING MAXIMUM ROUGHNESS

MAXIMUM LIMIT
PLACED ON TOP

SPECIFYING MINIMUM AND MAXIMUM ROUGHNESS

VALUES SHOWN ARE IN MICROINCHES

RECOMMENDED ROUGHNESS AVERAGE VALUES		N SERIES OF ROUGHNESS GRADE NUMBERS
MICROINCHES μ in.	MICROMETERS μ m	
2000	50	N12
1000	25	N11
500	12.5	N10
250	6.3	N9
125	3.2	N8
63	1.6	N7
32	0.8	N6
16	0.4	N5
8	0.2	N4
4	0.1	N3
2	0.05	N2
1	0.025	N1

FIGURE 22.7 Recommended roughness average ratings.

SUPPLEMENTAL INFORMATION

A table showing the production methods that may be used to obtain a required roughness average is shown in Figure 22.8. Roughness averages are shown in microinches, micrometers, and the ISO "N" series numbers. The system used on drawings will depend on whether the print is metric or the U.S. Customary system of dimensioning is used. You will note that there are overlapping choices that will produce the desired finish.

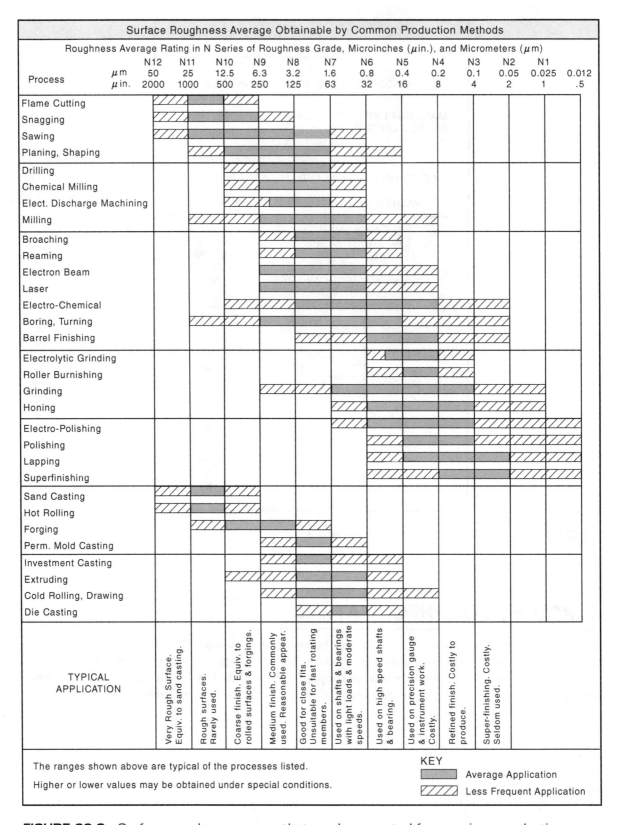

FIGURE 22.8 Surface roughness ranges that can be expected from various production methods.

Figure 22.9 shows examples of surface roughness ratings applied to a surface texture symbol. A further explanation regarding the characteristics of the finish produced at each rating. The roughness of the finish allowed increases as the number specified in micrometers and microinches increases.

MICROMETERS RATING	MICROINCHES RATING	APPLICATION
25	1000	Rough, low-grade surface resulting from sand casting, torch or saw cutting, chipping, or rough forging. Machine operations are not required because appearance is not objectionable. This surface, rarely specified, is suitable for unmachined clearance areas on rough construction items.
12.5	500	Rough, low-grade surface resulting from heavy cuts and coarse feeds in milling, turning, shaping, boring, rough filing, disc grinding, and snagging. It is suitable for clearance areas on machinery, jigs, and fixtures. Sand casting or rough forging produces this surface.
6.3	250	Coarse production surface, for unimportant clearance and cleanup operation, resulting from coarse surface grind, rough file, disc grind, rapid feeds in turning, milling, shaping, drilling, boring, grinding, etc., where tool marks are not objectionable. The natural surfaces of forgings, permanent mold castings, extrusions, and rolled surfaces also produce this roughness. It can be produced economically and is used on parts where stress requirements, appearance, and conditions of operations and design permit.
3.2	125	The roughest surface recommended for parts subject to loads, vibration, and high stress. It is also permitted for bearing surfaces when motion is slow and loads light or infrequent. It is a medium commercial machine finish produced by relatively high speeds and fine feeds taking light cuts with sharp tools. It may be economically produced on lathes, milling machines, shapers, grinders, etc., or on permanent mold castings, die castings, extrusion, and rolled surfaces.
1.6	63	A good machine finish produced under controlled conditions using relatively high speeds and fine feeds to take light cuts with sharp cuttings. It may be specified for close fits and used for all stressed parts, except fast rotating shafts, axles, and parts subject to severe vibration or extreme tension. It is satisfactory for bearing surfaces when motion is slow and loads light or infrequent. It may also be obtained on extrusions, rolled surfaces, die castings, and permanent mold casting when rigidly controlled.
0.8	32	A high-grade machine finish requiring close control when produced by lathes, shapers, milling machines, etc., but relatively easy to produce by centerless, cylindrical, or surface grinders. Also, extruding, rolling or die casting may produce a comparable surface when rigidly controlled. This surface may be specified in parts where stress concentration is present. It is used for bearings when motion is not continuous and loads are light. When finer finishes are specified, production costs rise rapidly; therefore, such finishes must be analyzed carefully.
0.4	16	A high-quality surface produced by fine cylindrical grinding, emery buffing, coarse honing, or lapping, it is specified where smoothness is of primary importance, such as rapidly rotating shaft bearings, heavily loaded bearing and extreme tension members.
0.2	8	A fine surface produced by honing, lapping, or buffing. It is specified where packings and rings must slide across the direction of the surface grain, maintaining or withstanding pressures, or for interior honed surface of hydraulic cylinders. It may also be required in precision gauges and instrument work, or sensitive value surfaces, or on rapidly rotating shafts and on bearings where lubrication is not dependable.
0.1	4	A costly refined surface produced by honing, lapping, and buffing. It is specified only when the design requirements make it mandatory. It is required in instrument work, gauge work, and where packing and rings must slide across the direction of surface grain such as on chrome plated piston rods, etc. where lubrication is not dependable.
0.05 0.025	2 1	Costly refined surfaces produced by only the finest of modern honing, lapping, buffing, and superfinishing equipment. These surfaces may have a satin or highly polished appearance depending on the finishing operation and material. These surfaces are specified only when design requirements make it mandatory. They are specified on fine or sensitive instrument parts or other laboratory items, and certain gauge surfaces, such as precision gauge blocks.

FIGURE 22.9 Description of various surface roughness ratings.

ASSIGNMENT D-17: PRESSURE PLUG

1. What does roughness refer to when measuring surface texture? _____

2. What material is the Pressure Plug made from? _____

3. How deep is the ∅ .62 hole? _____

4. What is the maximum allowable dimension for the 1.125 diameter? _____

5. What is the minimum allowable dimension for the 1.125 diameter? _____

6. What type of section is shown at A-A? _____

7. What is the outside diameter of the Pressure Plug? _____

8. Does the ∅ .12 hole go all the way through the Pressure Plug? _____

9. What roughness height is specified on the ∅ .8750 hole? _____

10. Is the surface finish on the outside diameter of the Pressure Plug smoother or rougher than the surface finish on the ∅ .8750 hole? _____

11. How deep is the ∅ .8750 hole? _____

12. What change was made at ①? _____

13. How much was the dimension at ② shortened after the change? _____

14. What tolerance is allowed on the ∅ .62 hole? _____

15. What unit of measure is a microinch? _____

16. What chamfer is required on the Pressure Plug? _____

17. What tolerance is allowed on the 3.375 length? _____

18. How many pressure plugs are required? _____

19. What is the dimension for distance ⊗? _____

20. What is the dimension for distance ⓨ? _____

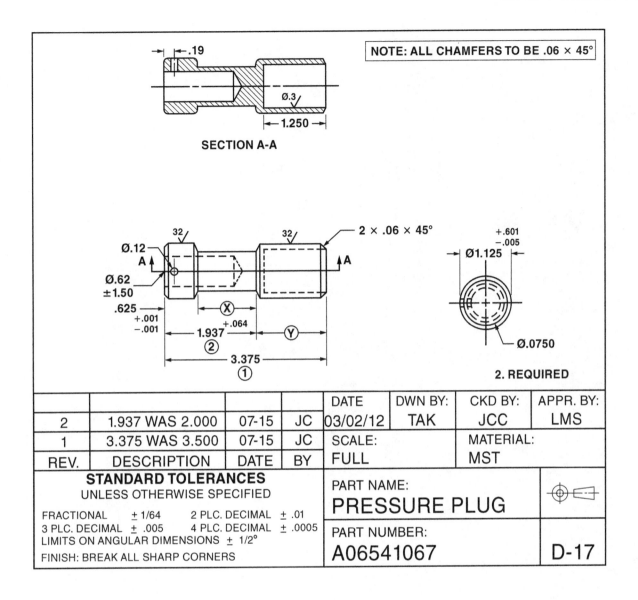

NOTE: ALL CHAMFERS TO BE .06 × 45°

SECTION A-A

2 × .06 × 45°

2. REQUIRED

2	1.937 WAS 2.000	07-15	JC	DATE	DWN BY:	CKD BY:	APPR. BY:
1	3.375 WAS 3.500	07-15	JC	03/02/12	TAK	JCC	LMS
REV.	DESCRIPTION	DATE	BY	SCALE: FULL		MATERIAL: MST	

STANDARD TOLERANCES UNLESS OTHERWISE SPECIFIED	PART NAME: PRESSURE PLUG	
FRACTIONAL ± 1/64 2 PLC. DECIMAL ± .01 3 PLC. DECIMAL ± .005 4 PLC. DECIMAL ± .0005 LIMITS ON ANGULAR DIMENSIONS ± 1/2° FINISH: BREAK ALL SHARP CORNERS	PART NUMBER: A06541067	D-17

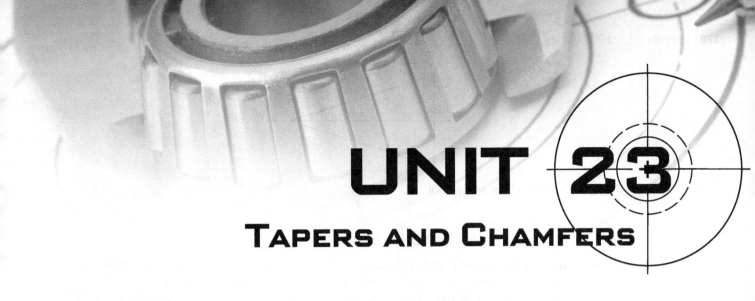

UNIT 23

TAPERS AND CHAMFERS

TAPERS

A *taper* is defined as a gradual and uniform increase or decrease in size along a given length of a part. Tapers may be conical or flat and are specified on engineering drawings in degrees of taper, a taper ratio, or by notation that a standard industrial taper is required.

Conical Tapers

A tapered surface on a round part is called a *conical taper*. Internal and external conical tapers are used extensively for alignment and holding purposes between mating parts. Machine spindles and tapered shank tools such as drills, reamers, mill cutters, drill chucks, and many other tools have standard tapers. The Brown and Sharpe, Morse, and American National Standard are the most common standard tapers used in industry. Figure 23.1 shows an example of a #3 American National Standard taper notation for a conical shaft.

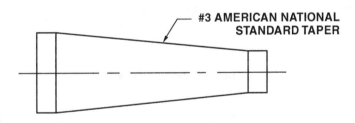

FIGURE 23.1   Standard taper.

Tapers that are nonstandard require dimensions for the large diameter and small diameter of the taper and the length of the tapered section. The rate of taper may be specified in degrees, taper per inch (TPI), or as a taper ratio.

When conical tapers are specified as a ratio, a conical taper symbol is used. The conical symbol is placed at the end of a leader line and the taper ratio is displayed to the right of the symbol, Figure 23.2. A taper ratio is defined as the rate of increase or decrease of a taper along the length of the taper. For example, a 1:10 ratio means a one-unit change in diameter per ten units of horizontal distance along the taper.

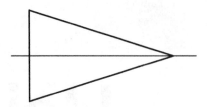

FIGURE 23.2 Conical taper symbol.

Determining Conical Taper Per Inch and Taper Ratio

If a ratio or angular dimension is not provided, the taper can be determined if the large diameter, small diameter, and length of the taper are known. For example, by subtracting the small diameter (d) from the large diameter (D) and dividing by the length of the taper (L), the amount of taper per inch or TPI can be determined. The TPI can then be converted to a ratio, Figure 23.3.

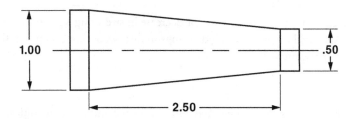

FIGURE 23.3 Calculating taper per inch and taper ratio.

EXAMPLE

Formula: $\dfrac{D - d}{L} = TPI$

Taper Per Inch $= \dfrac{1.00 - .50}{2.50} = \dfrac{.50}{2.50} = .2$ Taper Per Inch

Taper Ratio $= \dfrac{1.00}{.2} = 5$

Taper Ratio = 1:5 or 1.00 per 5.00 of taper length

Flat Tapers

Flat tapers are defined as a slope or inclined surface on a flat object. As with conical tapers, flat taper may be specified in degrees or as a ratio of the difference in heights at each end of the taper. When specified as a ratio, a flat taper symbol is used, Figure 23.4.

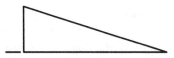

FIGURE 23.4 Flat taper symbol.

Examples of the methods used to specify tapers are shown in Figure 23.5.

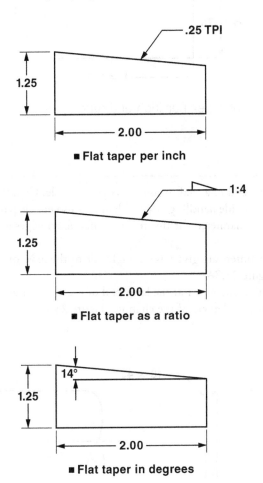

FIGURE 23.5 Methods of specifying flat tapers.

Determining Flat Taper Per Inch and Taper Ratio

If a flat taper ratio or taper per inch is not given, it can be determined using the same formula used to determine a conical taper. In order to use the formula, the height at both ends of the taper as well as the length of the taper must be known. Using the formula, subtract the height at the small end of the taper (h) from the height at the large end (H) and divide by the length of the taper to determine the TPI. The taper ratio can be determined using the TPI number. An example is provided in Figure 23.6, in which the dimensions on the part are H = .75, h = .50, and L = 2.00.

EXAMPLE

Formula: $\dfrac{D - d}{L} = TPI$ Taper Per Inch $= \dfrac{.75 - .50}{2.00} = \dfrac{.25}{2.00} = .125$ Taper Per Inch

Taper Ratio $= \dfrac{1.00}{.125} = 8$

Taper Ratio = 1:8

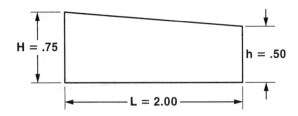

FIGURE 23.6 Calculating a flat taper per inch and taper ratio.

CHAMFERS

A **chamfer** is an angle cut on the end of a shaft or on the edge of a hole. Chamfers remove sharp edges, add to the appearance of the part, and provide handling safety. Chamfers also enable parts to be assembled more easily. The ends of screws and bolts are chamfered for this reason. Chamfered edges are usually cut at angles ranging from 30 degrees to 45 degrees.

Dimensions for external chamfers are given as an angle cut to the axis, or centerline, and a linear length. Two examples are shown in Figure 23.7A.

Internal chamfers are given as a diameter at the large end of the chamfer and an angular dimension on one side or as a diameter and the included angle of the chamfer, Figure 23.7B.

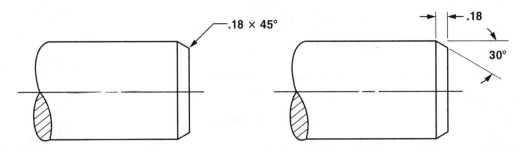

FIGURE 23.7A External chamfers.

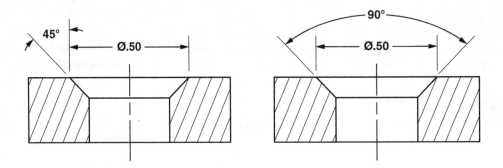

FIGURE 23.7B Internal chamfers.

ASSIGNMENT D-18: OFFSET BRACKET

1. What is the height of the shaft carrier? _____

2. What is the angle of the offset arm to the body of the piece? _____

3. What is the center-to-center measurement of the length of the offset arm? _____

4. What radius forms the upper end of the offset arm? _____

5. What is the width of the bolt slot in the body of the bracket? _____

6. What is the length, center-to-center, of the bolt slot? _____

7. What revision was made at ①? _____

8. What is the radius of the fillet between the pad and the body of the Offset Bracket? _____

9. What size chamfer is required on the Ø 10 hole? _____

10. What revision was made at ②? _____

11. What size oil hole is required in the shaft carrier? _____

12. How far is the oil hole from the finished face on the shaft carrier? _____

13. What tolerance is applied to the Ø 10 hole? _____

14. What is the diameter of the shaft carrier body? _____

15. Determine distance Ⓐ. _____

16. Determine distance Ⓑ. _____

17. Determine distance Ⓒ. _____

18. What surface finish is required on the shaft carrier and pad? _____

19. What does the line under the 136-mm dimension indicate? _____

20. Calculate the taper per inch and taper ratio for the following object.

    a. Taper Per Inch = _____

    b. Taper Ratio = ▷▢ _____

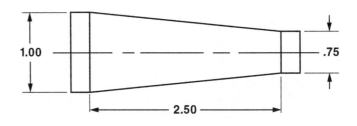

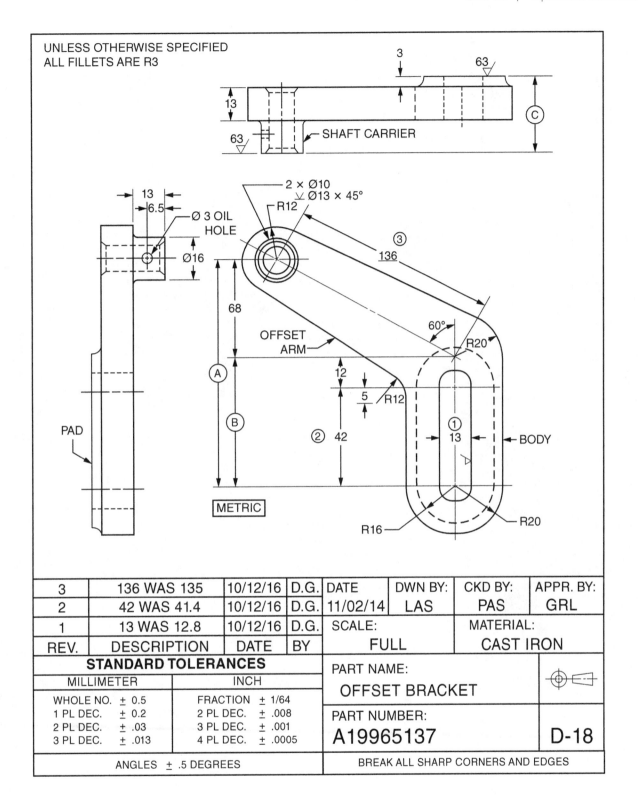

UNLESS OTHERWISE SPECIFIED
ALL FILLETS ARE R3

SHAFT CARRIER

13
63
3
63

C

13
6.5
Ø 3 OIL HOLE
Ø16

2 × Ø10
⌄ Ø13 × 45°
R12

③
136

68

OFFSET ARM

60°
R20

12
5
R12

②
42

①
13
BODY

A

B

PAD

METRIC

R16
R20

3	136 WAS 135	10/12/16	D.G.	DATE	DWN BY:	CKD BY:	APPR. BY:
2	42 WAS 41.4	10/12/16	D.G.	11/02/14	LAS	PAS	GRL
1	13 WAS 12.8	10/12/16	D.G.	SCALE:		MATERIAL:	
REV.	DESCRIPTION	DATE	BY	**FULL**		**CAST IRON**	

STANDARD TOLERANCES		PART NAME:	
MILLIMETER	INCH	**OFFSET BRACKET**	⊕◁
WHOLE NO. ± 0.5	FRACTION ± 1/64		
1 PL DEC. ± 0.2	2 PL DEC. ± .008	PART NUMBER:	
2 PL DEC. ± .03	3 PL DEC. ± .001	**A19965137**	**D-18**
3 PL DEC. ± .013	4 PL DEC. ± .0005		
ANGLES ± .5 DEGREES		BREAK ALL SHARP CORNERS AND EDGES	

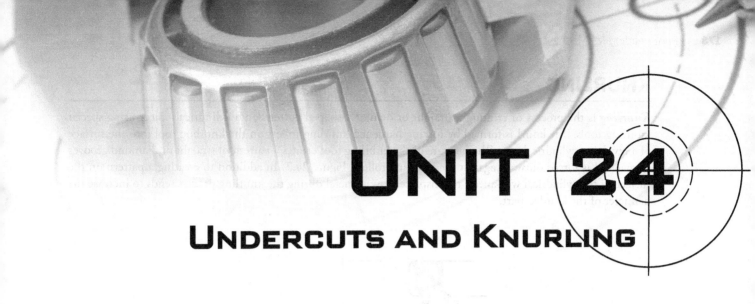

UNIT 24

UNDERCUTS AND KNURLING

UNDERCUTS

Undercutting, or *Necking*, as it is sometimes called, is the process of cutting a recess in a cylindrical part. Undercuts are often machined at the end of a threaded section or where a shaft changes diameter. Undercutting, or necking, operations remove any material that has been left from a prior machining process, thus allowing mating parts to seat flush against each other, Figure 24.1.

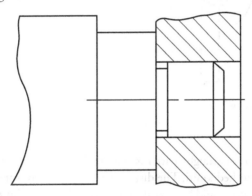

FIGURE 24.1   Part seated flush in a hole.

Dimensioning Undercuts

Necks are usually dimensioned with a leader and a note specifying the width of the undercut and the diameter or depth. The bottom of the undercut may be square or a radius may be specified to add strength and reduce the possibility of breakage. If a radius is required, it should be specified in the notation and the finished diameter measured at the bottom of the radius. Figure 25.2 shows two examples of dimensioning undercuts.

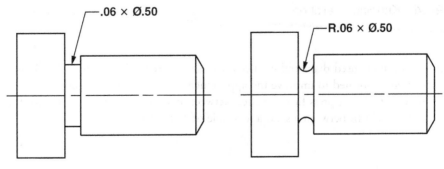

UNDERCUT WITH RADIUS

FIGURE 24.2   Dimensioning undercuts.

KNURLING

Knurling is the process of creating a straight or diamond-shaped pattern on a cylindrical object using special knurling tools. The knurl is formed by forcing hardened knurling rollers on the knurling tool into the surface of a revolving cylindrical part. The pressure of the knurling tool creates a pattern of straight or diamond grooves as material is forced outward against the knurling rollers, Figure 24.3. In addition to creating a pattern on the surface of the cylindrical workpiece, the displacement of metal during the knurling process tends to increase the diameter of the knurled part.

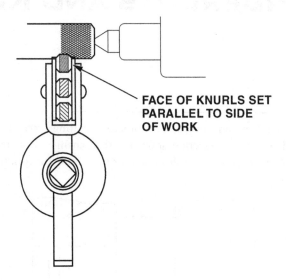

FIGURE 24.3 Knurling operation.

Knurls are made in diamond or straight patterns, Figure 24.4. The diametral pitch (DP) of the knurling tool determines the size of the finished knurl. The distance between the knurling grooves decreases as the diametral pitch increases. A 64DP knurl would be coarser than a 128DP knurl, for example.

FIGURE 24.4 Knurling operation.

The most commonly used diametral pitches for knurling are 64DP, 96DP, 128DP, and 160DP. Knurling operations are often performed to improve the appearance of a part, provide a gripping surface, or to increase the diameter of a part when a press fit is required between mating parts. Straight knurls, for example, may be specified to create a tight fit between a shaft and a hole of equal diameter.

Dimensioning Knurls

Knurling is generally dimensioned by specifying type, pitch, length of knurl, and diameter of the shaft before knurling, Figure 24.5. The finished diameter of the part after knurling may be specified if it is important to do so. If a straight knurling operation is being performed to create a tight fit in a hole, for example, the minimum diameter after knurling should be specified with a dimension and notation.

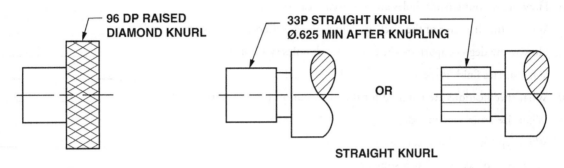

96 DP RAISED
DIAMOND KNURL

33P STRAIGHT KNURL
Ø.625 MIN AFTER KNURLING

OR

STRAIGHT KNURL

FIGURE 24.5 Dimensioning knurls.

ASSIGNMENT D-19: CAM CARRIER SUPPORT

1. What is the largest diameter of the cam carrier support?   _____

2. What is the overall length?   _____

3. What is the outside diameter of the hub?   _____

4. How long is the hub?   _____

5. Determine dimension Ⓕ.   _____

6. How many countersunk holes are in the cam carrier support?   _____

7. What is the diameter of the circle on which the countersunk holes are located?   _____

8. How many degrees apart are the countersunk holes spaced?   _____

9. What surface finish is specified on the ⌀ 1.004 hole?   _____

10. What specifications are required for the undercut on the ⌀ 1.400?   _____

11. What size knurl is required?   _____

12. What type of knurl is specified?   _____

13. What type of section view is shown?   _____

14. What material is specified in the title block?   _____

15. What surface finish is specified on the ⌀ 1.40 hub diameter?   _____

16. What is the largest size to which the hole through the center can be machined?   _____

17. What is the smallest size to which the hole through the center can be machined?   _____

18. What is the diameter of the recess into the ⌀ 2.75?   _____

19. What is the depth of the recess?   _____

20. What is the amount and degree of chamfer on the ⌀ 2.75?   _____

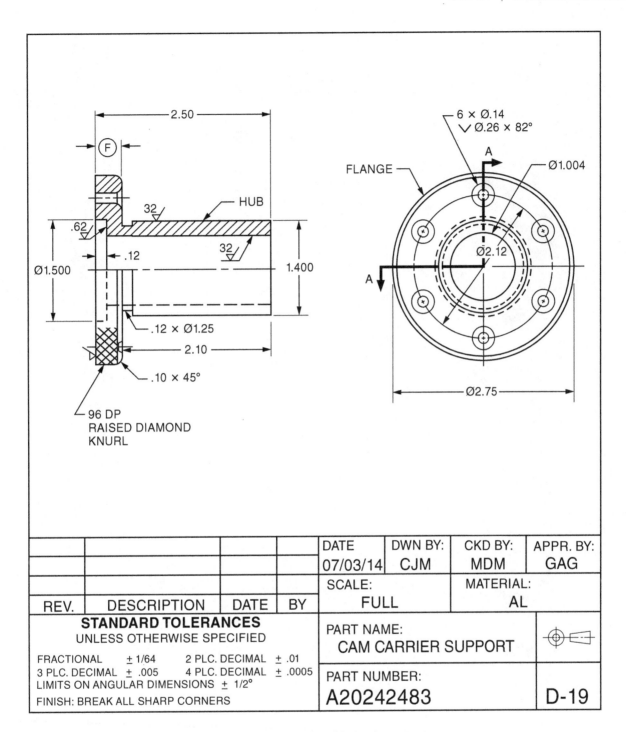

				DATE	DWN BY:	CKD BY:	APPR. BY:
				07/03/14	CJM	MDM	GAG
				SCALE:		MATERIAL:	
				FULL		AL	
REV.	DESCRIPTION	DATE	BY				

STANDARD TOLERANCES UNLESS OTHERWISE SPECIFIED	PART NAME: CAM CARRIER SUPPORT	
FRACTIONAL ± 1/64 2 PLC. DECIMAL ± .01 3 PLC. DECIMAL ± .005 4 PLC. DECIMAL ± .0005 LIMITS ON ANGULAR DIMENSIONS ± 1/2° FINISH: BREAK ALL SHARP CORNERS	PART NUMBER: A20242483	D-19

UNIT 25

SCREW THREAD
SPECIFICATION

Screw threads are used in a variety of mechanical applications. They may be used to transfer motion, transmit power, or to fasten parts together. Threaded fasteners such as nuts, bolts, and screws are perhaps the most common application of screw threads. Screw threads may be internal or external and are available in a variety of diameters, shapes, fits, and so on. Therefore, all information needed to fully understand the thread requirements must be specified on the drawing.

SCREW THREAD TERMINOLOGY

Screw Thread Definition. A ridge of uniform section in the form of a helix on the external or internal surface of a cylinder or hole, Figure 25.1.

Major or Nominal Diameter. The diameter of the thread as measured across the crests of an external thread and the root of a internal thread.

Minor Diameter. The smallest diameter of the thread of a screw or nut.

Pitch. The distance from a point on a screw thread to a corresponding point on the next thread.

Angle of Thread. The included angle of the thread.

Crest. The top of the major diameter of the thread.

Root. The bottom of the major diameter of the thread.

Thread Depth. The distance from the root to the crest of the thread.

Axis. The centerline running lengthwise through a screw.

Pitch Diameter. An imaginary cylinder that passes through the threads at a point where the width of the thread and the width of the groove are equal.

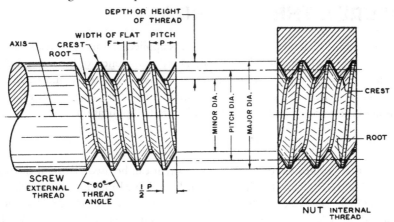

FIGURE 25.1 Screw thread terminology.

SCREW THREAD PROFILE OR FORM

Screw thread form refers to the shape of the screw thread as it appears in a profile view. A number of screw thread forms are used for various industrial applications. The form selected often depends on how the screw thread will be used, Figure 25.2.

Unified Inch Screw Thread Form.   The *Unified Inch Thread Form*, or Unified Thread Form as it is often called, has a 60-degree included thread angle and is the common thread form used in the United States, Canada, and Great Britain. The Unified Thread System is specified for most standard internal and external thread applications for holding, clamping, or assembly. It is also the screw thread system specified most often on the engineering drawings in this text.

Metric Screw Thread Form   The Unified Inch Thread served as the prototype for the Metric or M-Series thread form. Therefore, the profile of the Unified and the Metric threads is the same.

Buttress   *Buttress threads* are designed to handle extreme weight and to transmit power in one direction. They are capable of handling heavy lifting applications. A heavy duty automotive jack is a good example.

Acme   *Acme thread forms* are very strong and are used to transmit power and motion. A large flat at the crest and root are characteristics of Acme threads. Typical applications for Acme forms are lathe lead screws, vise screws, machine tables, and slides.

Worm   A *worm thread* is a special form of thread that is similar in shape to the Acme thread. In application, a cylindrical worm thread shaft engages with the threads on a mating part called a worm wheel. As the worm shaft rotates, it causes the worm wheel to rotate as well. Worm threads are used primarily to transmit power and motion.

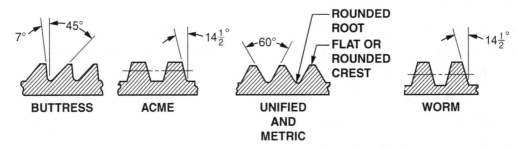

FIGURE 25.2   Thread profile.

STANDARD SCREW THREAD SERIES

Thread series refers to groups of diameter-pitch combinations applied to specific thread diameters. In simple terms, a thread series is a way of distinguishing threads based on the coarseness or fineness (pitch) of a thread. As stated previously, Unified inch screw threads are most frequently used for internal and external applications in the United States, Canada, and Great Britain. The Unified Inch Thread Series, along with ISO Metric Threads, are universally used around the world in manufacturing and everyday applications.

Unified Inch Screw Thread Series

Unified Inch Screw Threads are comprised of 11 thread series that can be used. Three of the series, the Unified Coarse (UNC), Unified Fine (UNF), and the Unified Extra-Fine (UNEF) have graded pitches, while the other eight series have constant pitches of 4-UN, 6-UN, 8-UN, 12-UN, 16-UN, 20-UN, 28-UN, and 32-UN threads per inch. In all examples, the UN is the abbreviation for Unified, while the number designation for the constant-pitch series is the number of threads per inch. Despite the numerous choices available, the preferred thread series for most general applications are the UNC and UNF.

Unified Coarse Series (UNC). Generally used for the majority of screw, bolt, and nut applications. The coarse thread allows for rapid assembly and disassembly, and is recommended for general use in machine construction.

Unified Fine Series (UNF). Recommended for applications where high strength is required. The Unified Fine thread has less thread depth than the Unified Coarse thread and is often used in situations where a hole may have a thin wall, such as a threaded hole in tubing.

Unified Extra-Fine Series (UNEF). Recommended where thin walled material is to be threaded and where depth of thread must be held to a minimum. For example, thin-walled tubing or thin nuts. They are also frequently used as adjusting screws where fine adjustment is required.

Constant-Pitch Thread Series. Used for special design applications where the same pitch is preferred regardless of the diameter of the threaded shaft of hole. The 8-Pitch Thread Series (8-UN), for example, has eight threads per inch for all diameters and is generally used on bolts for high-pressure pipe flanges, as well as cylinder head studs.

Metric M-Series Screw Thread Series

The ***Metric M-Series thread*** is used universally worldwide. Although the use of Unified threads in the United States remains at approximately 65 percent, the use of metric threads is increasing. The characteristics of Metric M-Series threads are the same as the Unified Inch Screw Thread, with one major difference. The dimensions and tolerances on Unified Inch Screw Threads are given in inch units while the Metric M-Series threads are specified in millimeters.

CLASSES OF FIT

Screw thread *fit* refers to the amount of allowance or tolerance between mating threads. There are three classes of fit that can be assigned to internal and external Metric M-Series and Unified Inch Threads. Each class of fit specifies the required tightness or looseness between threaded components.

The three classes of fit designations for Unified Inch Screw Threads are Class 1A, 2A, and 3A for external threads and Class 1B, 2B, and 3B for internal threads.

The class of fit designations for Metric threads are Tolerance Grade and Tolerance Position 8g, 6g, and 4g for external threads and Tolerance Grade and Tolerance Position 7H, 6H, and 5H for internal threads. The metric Tolerance Grade is indicated by a number. A number 6 is a standard fit. Numbers greater than 6 specify a looser fit and numbers lower than 6 specify a tighter fit.

Upper- and lowercase letters are used to identify the material limits for the crest and pitch diameters. An uppercase letter identifies the thread as internal, while a lowercase letter indicates that the thread is external. An upper- and lowercase H is standard for most applications.

Unified Inch Series - Class 1A for External Threads and Class 1B for Internal Threads.
Metric - Tolerance Grade and Position 8g for External Threads and 7H for Internal Threads.

- **Loose Fit** - Recommended only for screw thread work where clearance between mating parts is essential for rapid assembly and where shake or play is allowed.

Unified Inch Series - Class 2A for External Threads and Class 2B for Internal Threads.
Metric - Tolerance Grade and Position 6g for External Threads and 6H for Internal Threads.

- **Standard Fit**
Represents a high quality of commercial screw thread product. It is recommended for the great bulk of general screw thread applications.

Unified Inch Series - Class 3A for External Threads and Class 3B for Internal Threads.
Metric - Tolerance Grade and Position 4g for External Threads and 5H for Internal Threads.

- **Tight Fit**
Represents an exceptionally high grade of commercially threaded product. It is recommended only in cases where the high cost of precision tools and continual checking of tools and product is warranted.

UNIFIED INCH SCREW THREAD SPECIFICATION

A system of letters and numbers are used to specify screw threads on engineering drawings. The sequence followed for Unified Inch Screw threads is as follows and shown in Figure 25.3:

- Decimal nominal or outside diameter of the thread
- Number of threads per inch
- Thread form
- Thread series
- Class of fit
- Internal thread "A" or external thread "B" designation
- Left-Hand Thread—LH (Designation is not Required for Right-Hand Threads)

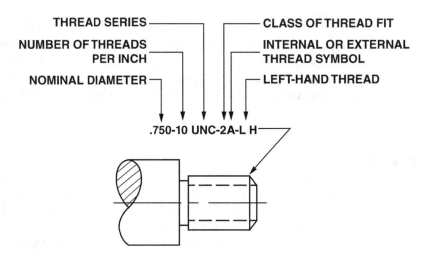

FIGURE 25.3   Unified inch thread specification.

METRIC SCREW THREAD SPECIFICATION

The order and method of Metric M-Series thread specifications differ somewhat from the Unified thread specifications. The dimensional requirements for tolerance grade and tolerance position may be difficult to understand for those having limited knowledge and experience with metric screw threads. The general explanation for grade and position tolerances was provided in the Class of Fit descriptions. The sequence for specifying Metric M-Series screw threads is as follows and shown in Figure 25.4:

- Metric symbol "M"
- Nominal diameter of the thread in millimeters
- Lower case "x"
- Pitch in millimeters
- Number and Letter—Specifies thread tolerance class designation for pitch diameter
- Number and Letter—Specifies thread tolerance class designation for major diameter of external thread or minor diameter for internal threads
- Left-Hand Thread—LH (Designation is not Required for Right-Hand Threads)

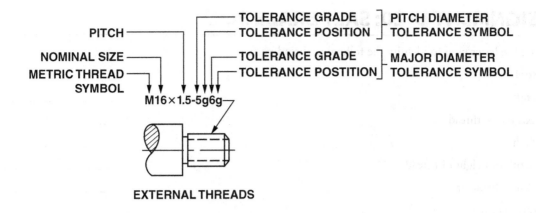

EXTERNAL THREADS

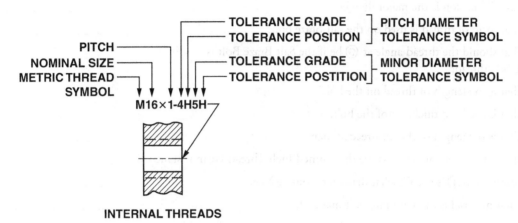

INTERNAL THREADS

FIGURE 25.4 Metric thread specification.

Specification of Left-Hand Threads

Threads may be cut as either a right-hand thread or a left-hand thread. When a right-hand thread is specified, no special symbol is put on the drawing. When a left-hand thread is required, the symbol LH is placed at the end of the thread specification for both Unified and Metric threads.

ASSIGNMENT D-20: SALT BRAZE BOLT

Indicate the letter used to identify the following thread parts:

1. Root _____

2. Crest _____

3. Axis of the thread _____

4. Pitch _____

5. Depth or height of thread _____

6. Minor diameter _____

7. Width of flat _____

8. What dimension is the major diameter? _____

9. What form of thread is used on the salt braze bolt? _____

10. What should the thread angle at Ⓖ be if the Salt Braze Bolt is
    a UNC thread? _____

11. What is the length of thread on the bolt? _____

12. What is the head thickness of the bolt? _____

13. What is the length of the unthreaded diameter? _____

14. Name two other countries where the Unified Inch Thread Form is standard. _____

15. If dimension Ⓕ is 0.187, what diameter would Ⓗ be? _____

16. What material is the salt braze bolt made of? _____

17. What thread form is used to transmit power in one direction and is
    capable of handling high stress? _____

18. Which thread form is similar to an Acme thread? _____

19. What unit of measure is used to specify the nominal diameter of
    a metric thread? _____

20. If the pitch of a screw thread is .125, what distance will the thread
    advance per revolution? _____

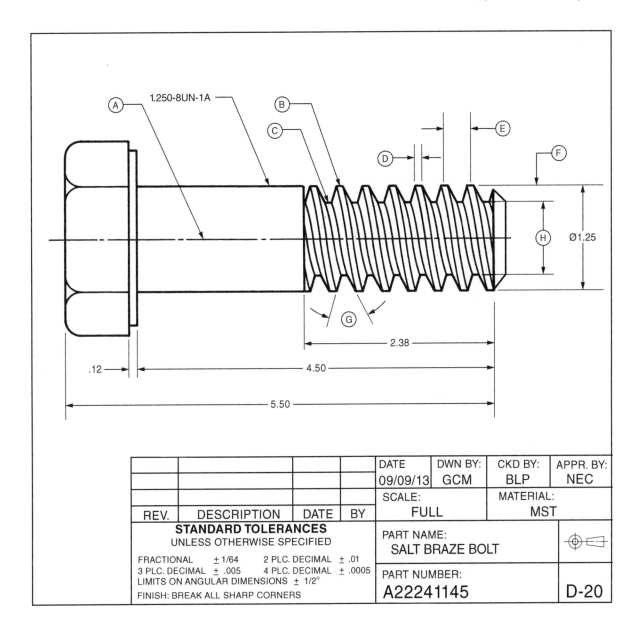

				DATE	DWN BY:	CKD BY:	APPR. BY:
				09/09/13	GCM	BLP	NEC
				SCALE:		MATERIAL:	
REV.	DESCRIPTION	DATE	BY	FULL		MST	

STANDARD TOLERANCES UNLESS OTHERWISE SPECIFIED	PART NAME: SALT BRAZE BOLT	⊕⊏
FRACTIONAL ± 1/64 2 PLC. DECIMAL ± .01 3 PLC. DECIMAL ± .005 4 PLC. DECIMAL ± .0005 LIMITS ON ANGULAR DIMENSIONS ± 1/2° FINISH: BREAK ALL SHARP CORNERS	PART NUMBER: A22241145	D-20

UNIT 26

SCREW THREAD
REPRESENTATION AND
THREADED FASTENERS

Internal and external threads can be represented on drawings in three different ways. They may be shown as a detailed, schematic, or simplified representation. Examples of an external threads drawn using each method are shown in Figure 26.1.

DETAILED REPRESENTATION

Pictorial or **detailed representations** show the thread form very close to how it actually appears. The detailed shape is very pleasing to the eye. However, the task of drawing pictorials is very difficult and time consuming. Therefore, pictorial representations are rarely used on threads if less than one inch in diameter, Figure 26.1A.

SCHEMATIC REPRESENTATION

A **schematic representation** does not show the true shape, or form, of the thread shape. Visible lines are used to show the nominal, or major, diameter of the thread. Crest and root lines are drawn perpendicular to the axis of the thread. Root lines are normally drawn thicker and darker than the crest lines. The spacing of the crest and root lines represent the approximate pitch of the thread, Figure 26.1B.

SIMPLIFIED REPRESENTATION

A **simplified representation** uses thread symbols to further reduce drafting time. A visible line is drawn to show the nominal diameter and two hidden lines drawn parallel to the axis of the thread represent the root, or depth, of the thread, Figure 26.1C.

EXTERNAL THREADS

An **external** or **profile view** of a threaded shaft clearly shows the method of representation used. However, when looking at an end view of the threaded diameter, the thread representation will appear the same regardless of the method selected. The crest diameter of the thread will be shown with a visible line, and the root diameter will be shown with a hidden line. If the end of the thread has been chamfered to the approximate thread depth, however, both the crest and root diameters may be shown as a solid line, Figure 26.1D.

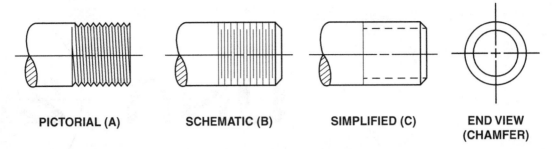

FIGURE 26.1   Methods of representing external screw threads.

INTERNAL THREADS

The examples provided in Figures 26.2A, B, and C show how a threaded hole will appear in a section view. Section views of threaded holes provide much more detail as to the thread representation method used. However, in an external profile view, as shown in Figure 26.2D, all three representations are the same. One pair of hidden lines are drawn parallel to the axis of the hole to represent the crest diameter and another pair of hidden lines represent the root diameter.

A front or end view of a threaded hole will also appear the same regardless of the representation method used, Figure 26.2E. The root diameter of the hole is shown as a visible line and the major, or nominal, diameter is shown with a hidden line. If a threaded hole is chamfered to the thread depth, the crest and root diameters may appear on the drawing as two concentric visible lines.

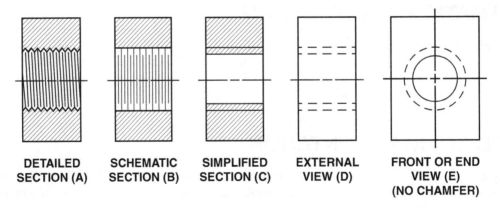

FIGURE 26.2   Methods of representing internal screw threads.

TAPPED HOLES

A *tapped hole* is a threaded hole that has been cut using a special tool called a *tap*. Taps are standard cutting tools available in English and metric sizes. Before threads can be cut, a hole must be machined using a tap drill of the appropriate size for the thread. Tapped holes may go all the way through a part or to a specified depth. Threads that are cut to a specified depth are often referred to as "blind holes" and may be shown in combination with the tap drill hole. The bottom of the tap drill hole is pointed to represent the drill point. Tapped holes are dimensioned with a leader and a depth symbol, Figure 26.3.

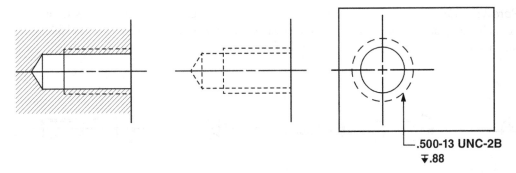

.500-13 UNC-2B
▼.88

FIGURE 26.3 Representing tapped holes.

THREADED FASTENERS

Threaded fasteners are used for many assembly, clamping, or adjusting purposes in machine work. Common threaded fasteners include a variety of screws, nuts, bolts, or studs. This unit provides an overview of some of the fastener hardware most frequently used in manufacturing and assembly applications. Other than on assembly drawings, threaded fasteners are seldom shown on engineering drawings. However, they may be specified as a notation on the print or on an assembly parts list. Therefore, it is important for the print reader to have a basic knowledge of threaded fasteners.

Screws

Screws are externally threaded fasteners used in assembly applications. Screws may be inserted through clearance holes in parts to be assembled and held in place by a nut, or they may be threaded into a threaded hole on a mating part. A variety of screw head and drive geometries are available for selection, depending on the fastening application.

Machine Screws. *Machine screws* are available in standard thread sizes and lengths. They are used frequently where small diameter fasteners are required for general assembly work. Machine screws are much like machine bolts or cap screws in appearance, Figure 26.4.

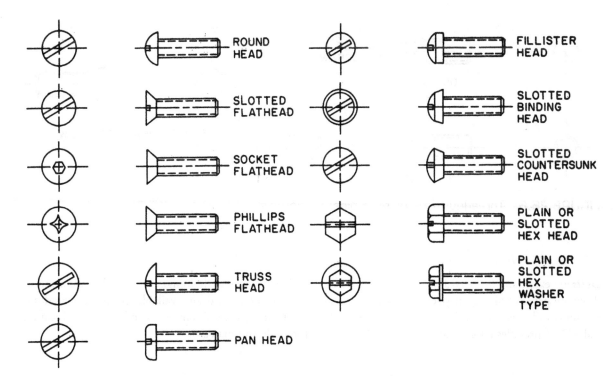

FIGURE 26.4 Threaded fastener examples: Machine screws.

Cap Screws. *Cap screws* are stronger and more precise than machine screws and are used for semi-permanent clamping or fastening. As with machine screws, cap screws are commercially available in many thread diameters, head geometries, and drive heads, Figure 26.5.

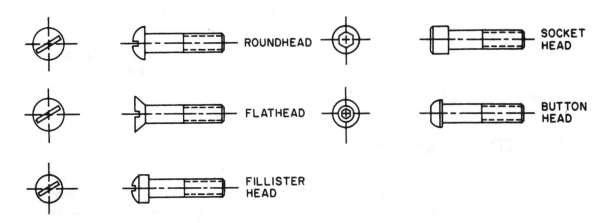

FIGURE 26.5 Threaded fastener examples: Cap screws.

Setscrews. *Setscrews* are used to prevent motion or slippage between two parts, such as pulleys or collars on a shaft. Setscrews are usually heat treated for added strength to resist wear. They may be either headed or headless, with a variety of point forms available. Some basic setscrews and point variations are shown in Figure 26.6.

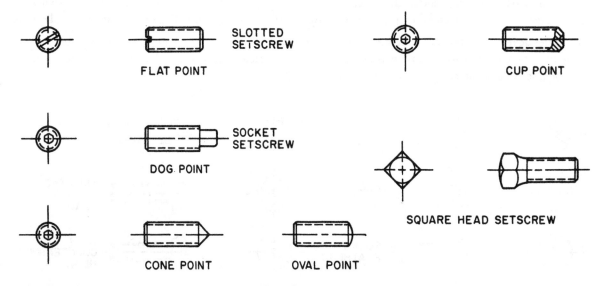

FIGURE 26.6 Threaded fastener examples: Setscrews.

Bolts and Nuts

Bolts are externally threaded fasteners that can be used in many clamping, holding, and assembly applications. They are generally larger in size than machine screws and cap screws and are available in different grades of hardness and tensile strength. Markings on the head of a bolt specify the bolt's material and grade classification. Table 26.1 provides information on bolt head markings for grade and material.

TABLE 26.1 Bolt and Screw Grade Markings

Grade Marking	Specification	Material
NO MARK	SAE — Grade I	Low or Medium Carbon Steel
	ASTM — A307	Low Carbon Steel
	SAE — Grade 2	Low or Medium Carbon Steel
	SAE — Grade 5	Medium Carbon Steel, Quenched and Tempered
	ASTM — A 449	
	SAE — Grade 5.2	Low Carbon Martensite Steel, Quenched and Tempered
A 325	ASTM — A 325 Type 1	Medium Carbon Steel, Quenched and Tempered Radial dashes optional
A 325	ASTM — A 325 Type 2	Low Carbon Martensite Steel, Quenched and Tempered
A 325	ASTM — A 325 Type 3	Atmospheric Corrosion (Weathering) Steel, Quenched and Tempered
BC	ASTM — A 354 Grade BC	Alloy Steel, Quenched and Tempered
	SAE — Grade 7	Medium Carbon Alloy Steel, Quenched and Tempered, Roll Threaded After Heat Treatment
	SAE — Grade 8	Medium Carbon Alloy Steel, Quenched and Tempered
	ASTM — A 354 Grade BD	Alloy Steel, Quenched and Tempered
	SAE — Grade 8.2	Low Carbon Martensite Steel, Quenched and Tempered
A 490	ASTM — A 490 Type 1	Alloy Steel, Quenched and Tempered
A 490	ASTM — A 490 Type 3	Atmospheric Corrosion (Weathering) Steel, Quenched and Tempered

Machine Bolts. *Machine bolts* are used to clamp or hold two or more parts together. They are not as precise as cap screws and are not available in as large a variety of head forms. Machine bolts have either square or hex-shaped heads, Figure 26.7.

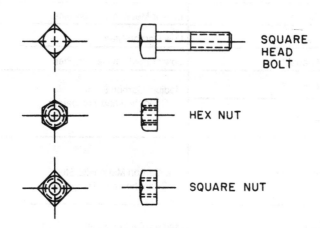

FIGURE 26.7 Threaded fastener examples: Bolts.

Nuts. Various types of *nuts* are commercially available for use on assemblies. The designer must select the proper style best suited for each application. Some typical styles of nuts include square, hex, jam, castle, and flange, Figure 26.8.

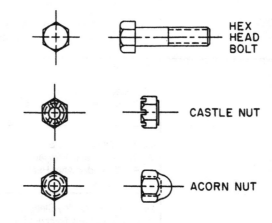

FIGURE 26.8 Threaded fastener examples: Nuts.

Studs or Stud Bolts. *Studs* or *Stud bolts* are headless bolts with threads on each end. One end can be threaded into a mating part and held in place with a nut, or a combination of nut and washer. In some applications, a nut and washer may be fastened to the opposite ends of the stud. Studs are frequently used in clamping applications such as securing a part to a machining table. They are also commonly used to secure equipment to a floor or base, Figure 26.9.

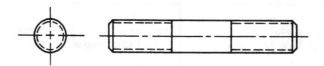

FIGURE 26.9 Threaded fastener examples: Stud bolt.

Washers

Washers are often used in combination with screws, nuts, bolts, and studs. Washers help to distribute clamping pressure over a wider area. They also prevent surface marring that may result from contact with the head of the bolt or screw.

Common Washers. The most common types of machine washers are the ***plain flat washer*** and the ***spring lock washer***. Flat washers provide a larger load bearing surface for the head of the bolt.

Spring lock washers prevent ***backing off*** or loosening of threaded fastener assemblies. Appendix Table 12 provides information on flat washers and spring lock washers.

THREADED FASTENER ASSEMBLIES

Figure 26.10 illustrates several typical threaded fastener assembly applications using screws, bolts, nuts, and washers. In many applications, the screw heads are recessed into the assembly by countersinking or counterboring.

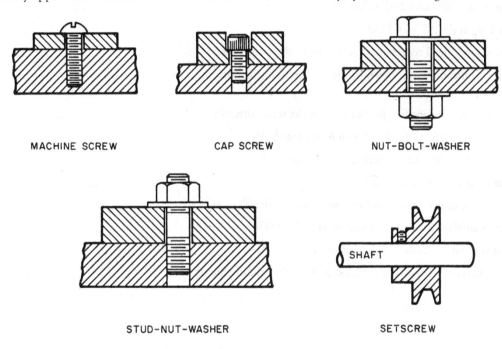

FIGURE 26.10 Typical fastener assemblies.

THREADED FASTENER SIZE

The size of a threaded fastener such as a screw, bolt, or stud is determined by the nominal thread diameter. The body is used to designate the dimension of length. However, the thread length may vary with the diameter or style of fastener, Figure 26.11.

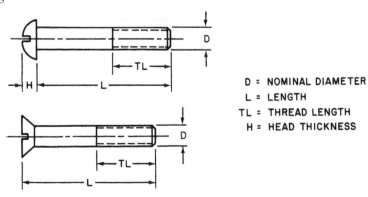

D = NOMINAL DIAMETER
L = LENGTH
TL = THREAD LENGTH
H = HEAD THICKNESS

FIGURE 26.11 Fastener size specifications.

ASSIGNMENT D-21: SPINDLE SHAFT

1. What is the overall length of the Spindle Shaft? _____

2. What revision was made at ②? _____

3. What is the largest diameter of the shaft? _____

4. How many diameters are threaded? _____

5. What are the specifications for the left-handed thread? _____

6. How many threads per inch are required for the 1.000-12 UNF-3A thread? _____

7. What thread series is specified for the threads? _____

8. What class of fit is specified on the threads? _____

9. Is the class of fit a standard or loose fit? _____

10. How long is the .875-14 UNF-3A thread? _____

11. What screw thread representation is used to show the screw threads? _____

12. What specifications are shown for the undercut at the bottom of the 1.00 inch thread? _____

13. What chamfer is specified for the ends of the screw threads? _____

14. What is the width of the flat shown in section A-A? _____

15. What are the dimensions specified for the keyseat? _____

16. What revision was made at ①? _____

17. What is the upper limit dimension specified for the ⌀ 1.500? _____

18. What is smallest diameter allowed on the ⌀ 1.000? _____

19. What standard tolerance is allowed on the ⌀ 2.12? _____

20. What operation is specified for both ends of the Spindle Shaft? _____

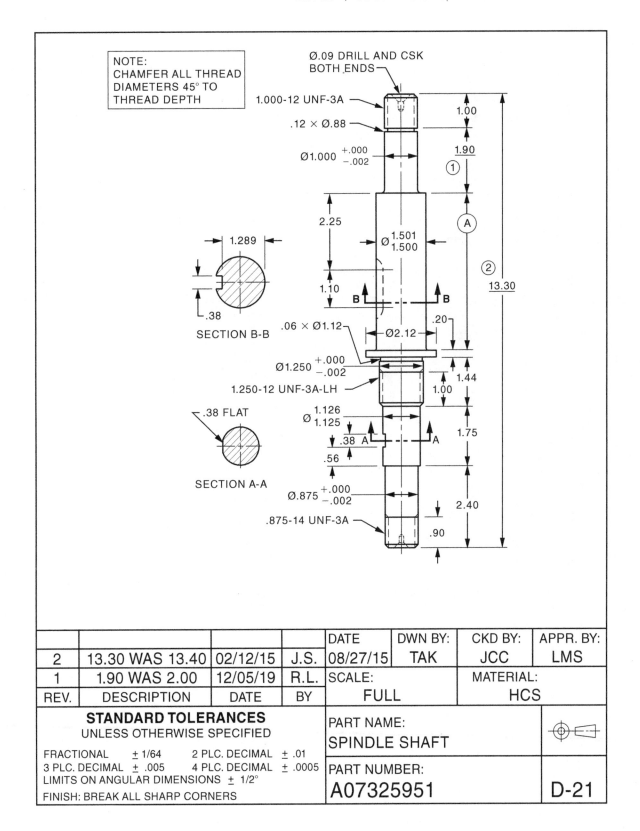

NOTE:
CHAMFER ALL THREAD
DIAMETERS 45° TO
THREAD DEPTH

Ø.09 DRILL AND CSK
BOTH ENDS

1.000-12 UNF-3A

.12 × Ø.88

Ø1.000 +.000 / −.002

1.00

1.90

①

2.25

Ø 1.501 / 1.500

Ⓐ

②

13.30

1.289

1.10

B   B

.38

SECTION B-B

.06 × Ø1.12

Ø2.12

.20

Ø1.250 +.000 / −.002

1.250-12 UNF-3A-LH

1.44

1.00

.38 FLAT

Ø 1.126 / 1.125

.38 A

A

1.75

SECTION A-A

.56

Ø.875 +.000 / −.002

2.40

.875-14 UNF-3A

.90

2	13.30 WAS 13.40	02/12/15	J.S.	DATE 08/27/15	DWN BY: TAK	CKD BY: JCC	APPR. BY: LMS
1	1.90 WAS 2.00	12/05/19	R.L.	SCALE: FULL		MATERIAL: HCS	
REV.	DESCRIPTION	DATE	BY				

STANDARD TOLERANCES
UNLESS OTHERWISE SPECIFIED

FRACTIONAL   ± 1/64   2 PLC. DECIMAL ± .01
3 PLC. DECIMAL ± .005   4 PLC. DECIMAL ± .0005
LIMITS ON ANGULAR DIMENSIONS ± 1/2°
FINISH: BREAK ALL SHARP CORNERS

PART NAME:
SPINDLE SHAFT

PART NUMBER:
A07325951

D-21

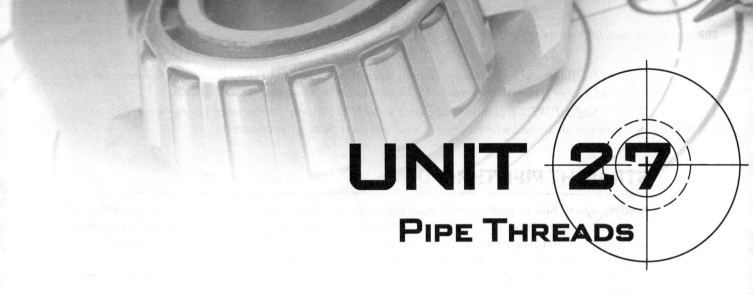

UNIT 27

PIPE THREADS

AMERICAN NATIONAL STANDARD PIPE THREADS

Pipe threads are used in many industrial applications where a pressure-tight joint or mechanical connection is required. Various styles of pipe threads are commercially available for use.

Two types of pipe threads are approved as American National Standard, or National Standard as they are often referred to. They are tapered pipe threads, designated by the initials NPT, and straight pipe threads, designated as NPS. Both thread series have the same 60-degree thread angle and the same pitch. A variety of tapered and straight pipe sizes are available, depending on the design function of the joint requirements. Some applications require a pressure-tight joint for sealing, while other applications may need a mechanical joint.

Pressure-Tight Pipe Joints

Pressure-tight joints may fall into one of the two categories:

1. Joints that provide gas or liquid pressure tightness when assembled with a sealer.
2. Joints that provide gas or liquid pressure tightness when assembled without a sealer. This type of pipe thread assembly is called a *dryseal pressure-tight joint*. Dryseal threads are used in automotive, aircraft, and marine applications.

Mechanical Joints

Mechanical joints are threaded assemblies that are not pressure-tight. Mechanical joints may be rigid or loose. Rigid joints are used for rail fittings. Loose joints are used on fixture assemblies and hose coupling connections.

TAPERED PIPE THREADS

Tapered pipe threads are recommended for general use. They have a standard taper per foot (TPF) of .750 and a thread angle of 60 degrees. The taper of the thread ensures easy starting and a tight joint when assembled. Tapered pipe threads are specified on engineering drawings with the letters NPT (National Pipe Thread).

The following system of letters is used to designate American National Standard taper pipe threads and common applications:

NPT—The most common commercially available thread form for pipe, pipe fittings, and valves. NPT thread forms can be used for mechanical joints or pressure-tight joints capable of preventing liquid or gas leakage when a sealer is applied.

NPTR—This thread is used for applications where a rigid mechanical railing joint is required.

NPTF—This thread is used for applications where it is desirable to have a pressure-tight joint without the use of a sealer applied to the threads. NPTF pipe forms are known as "dryseal" threads.

PTS-SAE SHORT—This thread is the same as the NPTF except it is shortened by one thread. PTS-SAE short threads are used where a dryseal thread and extra clearance are required.

STRAIGHT PIPE THREADS

Straight pipe threads are parallel to the axis of the thread. The thread form is the same as the American National Standard taper pipe thread. The number of threads per inch, thread angle, and thread depth is the same as the tapered version. Straight pipe thread fittings may be used for pressure-tight joints or mechanical joints. However, straight pipe pressure-tight fittings should only be used in low-pressure situations.

The following system of letters is used to designate American National straight threads and common applications.

NPSC—This type of thread is used on couplings where a low-pressure-tight joint is required.

NPSM—Used in loose fitting mechanical joint applications where no internal pressures exist.

NPSL—This type of thread is used for loose fitting mechanical joints where the maximum diameter of pipe is required.

NPSF—This thread is a dryseal thread and is used for internal applications only. NPSF threads are generally cut in soft or ductile material and used for fuel lines.

NPSI—This thread is also a dryseal thread used for internal applications. It is similar to the NPSF thread, but is slightly larger in diameter and is used in hard or brittle materials.

REPRESENTATION OF PIPE THREADS

Pipe threads are usually represented on drawings in schematic or simplified form. Views of internal and external pipe threads are drawn the same as the Unified Threads discussed in the previous units. The taper of the pipe thread is not normally shown unless needed. When a view of a tapered pipe thread is provided on a drawing, the amount of taper is generally exaggerated to provide clarity, Figure 27.1.

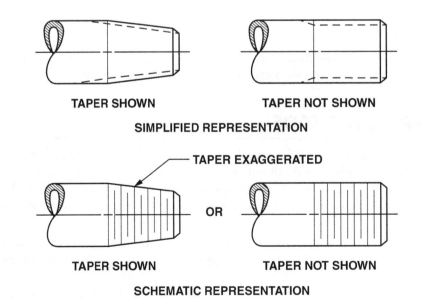

TAPER SHOWN   **TAPER NOT SHOWN**

SIMPLIFIED REPRESENTATION

TAPER EXAGGERATED

OR

TAPER SHOWN   **TAPER NOT SHOWN**

SCHEMATIC REPRESENTATION

FIGURE 27.1 Typical pipe thread representation.

SPECIFICATION OF PIPE THREADS

American National Standard pipe threads are specified on drawings using a sequence of numbers and letters. The thread specification should include the nominal size, the number of threads per inch, the thread series, and thread form. Additional information such as the minimum length of full thread engagement and dimensions for a countersink or chamfer may also be included in symbol or note form on the drawing. Thread specifications are called out on a print using a leader line and note, Figure 27.2.

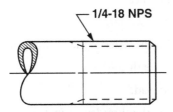

1/4-18 NPS

STRAIGHT PIPE SPECIFICATION

FIGURE 27.2 Typical pipe thread specifications.

ASSIGNMENT D-22: STUFFING BOX

1. How many threaded holes are required on the Stuffing Box?   _____

2. What is the distance between these holes?   _____

3. Are these threads cut all the way through the flange on the Stuffing Box?   _____

4. What are the specifications for the threaded holes?   _____

5. What does the NPS on the pipe thread mean?   _____

6. What application would this thread be used for?   _____

7. What thread is required on the outside diameter of the Stuffing Box?   _____

8. What does the 16 UN tell you about the thread?   _____

9. What is the length specified for this thread?   _____

10. What is the fillet size specified between the threaded diameter
    and the flange?   _____

11. How thick is the flange?   _____

12. How long is the flange?   _____

13. What is the overall height of the Stuffing Box?   _____

14. What is the smallest diameter the hole through the center of the
    Stuffing Box can be bored?   _____

15. What is the largest diameter the hole through the center of the
    Stuffing Box can be bored?   _____

16. What is height of the $\varnothing$ 2.000?   _____

17. What tolerance is allowed on the length of the $\varnothing$ 2.000?   _____

18. What is the angle and depth of the chamfer shown on the $\varnothing$ 1.250 hole?   _____

19. What surface finish is specified on the $\varnothing$ 1.250?   _____

20. What material is the Stuffing Box made from?   _____

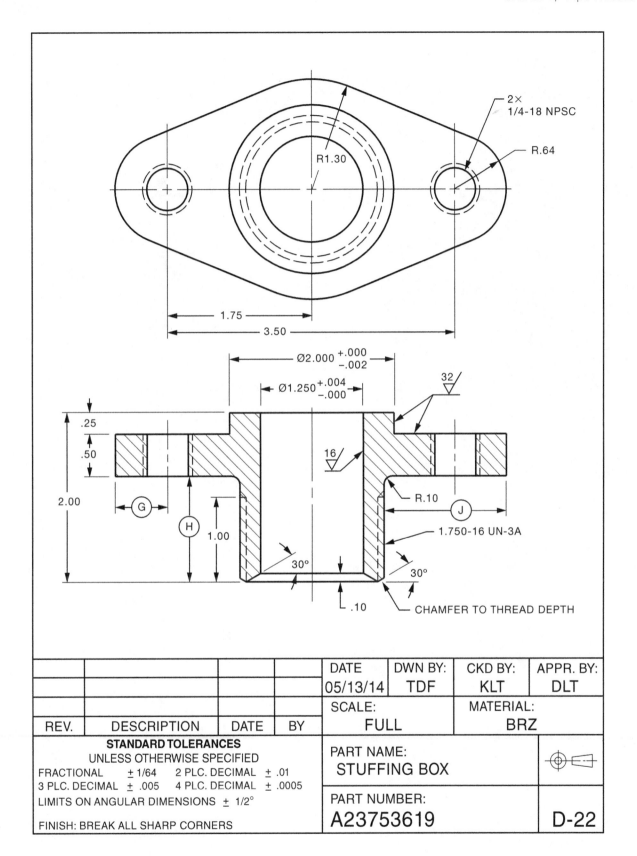

				DATE 05/13/14	DWN BY: TDF	CKD BY: KLT	APPR. BY: DLT
				SCALE: **FULL**		MATERIAL: **BRZ**	
REV.	DESCRIPTION	DATE	BY				

STANDARD TOLERANCES
UNLESS OTHERWISE SPECIFIED
FRACTIONAL ± 1/64   2 PLC. DECIMAL ± .01
3 PLC. DECIMAL ± .005   4 PLC. DECIMAL ± .0005
LIMITS ON ANGULAR DIMENSIONS ± 1/2°

FINISH: BREAK ALL SHARP CORNERS

PART NAME:
STUFFING BOX

PART NUMBER:
A23753619

D-22

UNIT 28
ASSEMBLY DRAWINGS

ASSEMBLY DRAWINGS

Industrial drawings often show two or more parts that must be put together to form an *assembly*. An *assembly drawing* shows the parts or details of a machine or structure in their relative positions as they appear in a completed unit, Figure 28.1.

In addition to showing how the parts fit together, the assembly drawing is used to:

- Represent the proper working relationship of the mating parts and the function of each.
- Provide a visual image of how the finished product should look when assembled.
- Provide overall assembly dimensions and center distances.
- Provide a bill of materials for machined or purchased parts required in the assembly.
- Supply illustrations that may be used for catalogs.

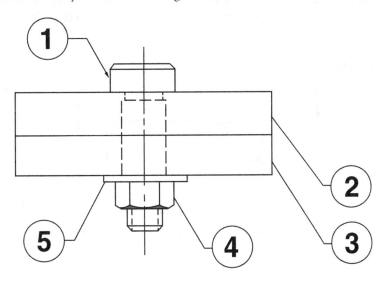

FIGURE 28.1   Assembly drawing.

SUB-ASSEMBLY DRAWINGS

Many large or complex machines are composed of smaller units that will become part of the final assembly of a machine or product. These sub-assemblies, as they are called, may consist of machined components, purchased parts, or both. Additionally, a sub-assembly may be a unit that is built by an outside supplier or provided by a different manufacturing facility within the same company. *Sub-assembly drawings* generally provide more detail and are usually shown on separate prints.

BILL OF MATERIALS OR PARTS LIST

Some information regarding a parts list or bill of materials was previously provided in Unit 2: Title Blocks, but is worth repeating here. The terms *bill of materials* or *parts list* are used interchangeably and refer to a section of the assembly drawings that lists all parts required for a completed assembly. A parts list may appear in a corner of the primary assembly drawing or on a separate sheet. Each item listed in the bill of materials is referred to as a detail. Non-standard parts may require a detail drawing, Figure 28.2.

5	1	KNURLED NUT	1 1/4 × 5/8	MST
4	1	SLIDE SHAFT	5/8 × 3 1/8	MST
3	1	SET SCREW	1/4 – 20 × 3/8	STD
2	1	V-ANVIL	5/8 × 2 1/16	MST
1	1	BASE	CASTING	CI
DET	REQ'D	DESCRIPTION	STOCK SIZE	MAT'L

FIGURE 28.2 Bill of materials or parts list.

DETAIL DRAWINGS

Details may be standard purchased parts such as machine screws, bolts, washers, springs, or nonstandard parts that must be manufactured or fabricated. Unaltered purchased parts do not require a detail drawing. The specifications for standard units are provided in the parts list or bill of materials.

Nonstandard parts require drawings that may appear on one sheet or on separate sheets. These *detail drawings* supply more specific information than is provided on the assembly drawing. All views, dimensions, and notes required to describe the part completely appear on the detail drawing, Figure 28.3.

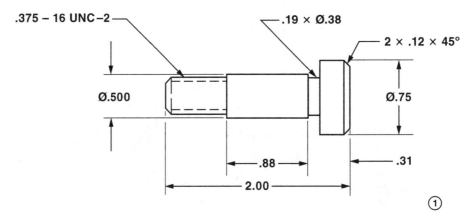

FIGURE 28.3 Detail drawing.

PARTS IDENTIFICATION SYMBOLS

The details of a mechanism are identified on an assembly with reference letters or numbers. These letters or numbers are contained in circles, or balloons, with leaders running to the part to which each refers, Figure 28.4. These symbols are also included in a parts list that gives a descriptive title for each part.

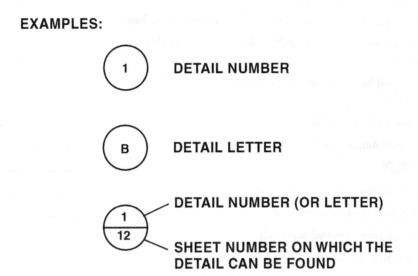

EXAMPLES:

1   **DETAIL NUMBER**

B   **DETAIL LETTER**

1 / 12   **DETAIL NUMBER (OR LETTER)**

      SHEET NUMBER ON WHICH THE DETAIL CAN BE FOUND

FIGURE 28.4 Detail part identification symbols.

DIMENSIONING

Assembly drawings should not be overloaded with dimensions that may be confusing to the print reader. Specific dimensional information should be provided on the detail drawings. Only such dimensions as center distances, overall dimensions, and dimensions that show the relationship of details to the assembly as a whole should be included. However, there are times when a simple assembly may be dimensioned so that no detail drawings are needed. In such cases, the assembly drawing becomes a working assembly drawing.

ASSIGNMENT D-23: MILLING JACK

1. How many details make up the jack assembly   _____

2. What is the thickness of the Ø 3.00 section of the base, not including the boss?   _____

3. What is the diameter of the boss on the jack base?   _____

4. What is the distance from the top of the boss to the top of the base?   _____

5. How far is the centerline of the Ø .625 hole from the centerline of the Ø .66 hole?   _____

6. What is the overall height of the assembled jack when it is in its lowest position?   _____

7. What size is the hole in the slide shaft?   _____

8. What is the detail number of the slide shaft?   _____

9. What is detail ③?   _____

10. How many sheets make up the drawing set?   _____

11. What material is the jack base made of?   _____

12. Of what material is the knurled nut made?   _____

13. What is the rough-cut stock size of the slide shaft?   _____

14. What is the maximum allowable diameter of the jack base?   _____

15. What is the size of the tapped hole in the jack base?   _____

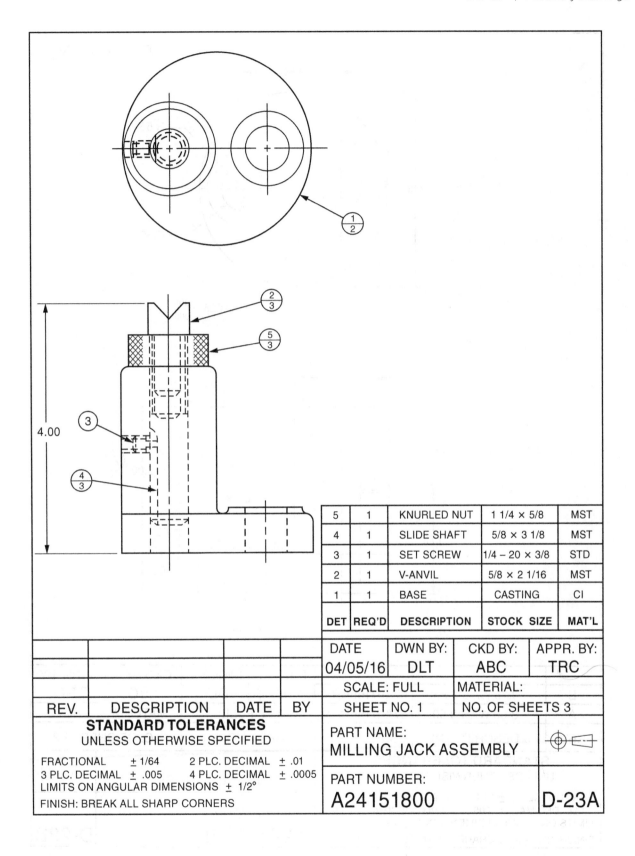

5	1	KNURLED NUT	1 1/4 × 5/8	MST
4	1	SLIDE SHAFT	5/8 × 3 1/8	MST
3	1	SET SCREW	1/4 – 20 × 3/8	STD
2	1	V-ANVIL	5/8 × 2 1/16	MST
1	1	BASE	CASTING	CI
DET	REQ'D	DESCRIPTION	STOCK SIZE	MAT'L

DATE	DWN BY:	CKD BY:	APPR. BY:
04/05/16	DLT	ABC	TRC
SCALE: FULL		MATERIAL:	
SHEET NO. 1		NO. OF SHEETS 3	

REV.	DESCRIPTION	DATE	BY

STANDARD TOLERANCES
UNLESS OTHERWISE SPECIFIED

FRACTIONAL ± 1/64 2 PLC. DECIMAL ± .01
3 PLC. DECIMAL ± .005 4 PLC. DECIMAL ± .0005
LIMITS ON ANGULAR DIMENSIONS ± 1/2°
FINISH: BREAK ALL SHARP CORNERS

PART NAME:
MILLING JACK ASSEMBLY

PART NUMBER:
A24151800   D-23A

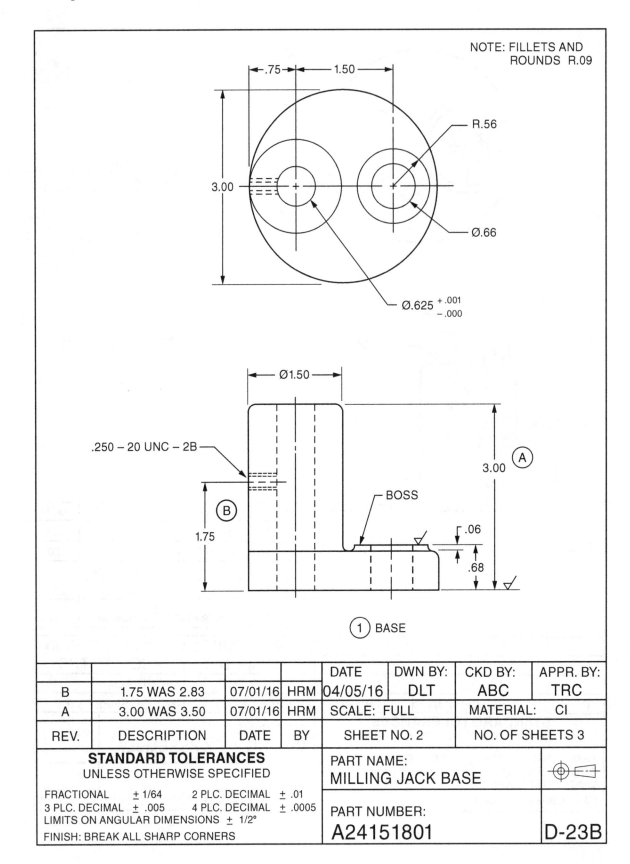

NOTE: FILLETS AND ROUNDS R.09

R.56

Ø.66

Ø.625 +.001 / −.000

.75

1.50

3.00

Ø1.50

.250 – 20 UNC – 2B

B

1.75

BOSS

.06

.68

3.00 Ⓐ

① BASE

				DATE	DWN BY:	CKD BY:	APPR. BY:
B	1.75 WAS 2.83	07/01/16	HRM	04/05/16	DLT	ABC	TRC
A	3.00 WAS 3.50	07/01/16	HRM	SCALE: FULL		MATERIAL: CI	
REV.	DESCRIPTION	DATE	BY	SHEET NO. 2		NO. OF SHEETS 3	

STANDARD TOLERANCES UNLESS OTHERWISE SPECIFIED	PART NAME: MILLING JACK BASE	⊕⟊
FRACTIONAL ± 1/64 2 PLC. DECIMAL ± .01 3 PLC. DECIMAL ± .005 4 PLC. DECIMAL ± .0005 LIMITS ON ANGULAR DIMENSIONS ± 1/2° FINISH: BREAK ALL SHARP CORNERS	PART NUMBER: A24151801	D-23B

V Anvil

1. What size chamfer is required on the anvil?   _____

2. How long is the Ø .374 diameter?   _____

3. How deep is the "V"?   _____

4. What are the dimensions for the neck?   _____

5. What is the largest diameter for the anvil?   _____

6. What is the upper limit dimension for the Ø .374 diameter?   _____

7. What tolerance is allowed on the 45° angle?   _____

8. What is the overall length of the part?   _____

9. How many V anvils are required?   _____

10. What detail number is the anvil?   _____

Slide Shaft

1. What is the depth of the Ø .375 hole?   _____

2. How long is the .625-18 UNF-2A thread?   _____

3. What type section is shown at AA?   _____

4. What size is the keyseat on the shaft?   _____

5. How long is the shaft keyseat?   _____

6. What class fit is required on the .625-18 UNF-2a threaded section? _____

7. What type of line is shown at Ⓐ?   _____

8. What method of thread representation is shown in the front view? _____

9. The section lining in section AA indicates what type of material? _____

10. What is the lower limit dimension for the Ø .625 diameter?   _____

Knurled Nut

1. What size knurl is required on the nut?   _____

2. How many finished surfaces are required?   _____

3. What is the thickness of the nut?   _____

4. What size tapped hole is required?   _____

5. What is the diameter of the knurled nut?   _____

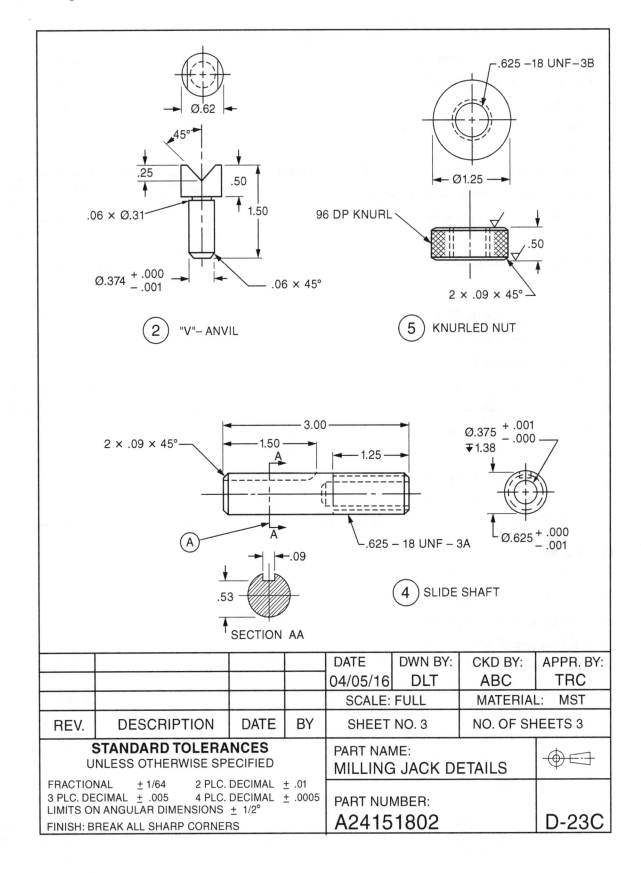

Ø.62

45°

.25

.50

1.50

.06 × Ø.31

Ø.374 +.000 / −.001

.06 × 45°

2   "V"– ANVIL

.625 –18 UNF–3B

Ø1.25

96 DP KNURL

.50

2 × .09 × 45°

5   KNURLED NUT

2 × .09 × 45°

3.00

1.50

1.25

A

A

A

.09

.53

SECTION AA

Ø.375 +.001 / −.000
⌴1.38

Ø.625 +.000 / −.001

.625 – 18 UNF – 3A

4   SLIDE SHAFT

				DATE	DWN BY:	CKD BY:	APPR. BY:
				04/05/16	DLT	ABC	TRC
				SCALE: FULL		MATERIAL: MST	
REV.	DESCRIPTION	DATE	BY	SHEET NO. 3		NO. OF SHEETS 3	

STANDARD TOLERANCES
UNLESS OTHERWISE SPECIFIED

FRACTIONAL   ± 1/64   2 PLC. DECIMAL ± .01
3 PLC. DECIMAL ± .005   4 PLC. DECIMAL ± .0005
LIMITS ON ANGULAR DIMENSIONS ± 1/2°
FINISH: BREAK ALL SHARP CORNERS

PART NAME:
MILLING JACK DETAILS

PART NUMBER:
A24151802

D-23C

UNIT 29
WELDING SYMBOLS

Welding is a process used for permanently joining parts together. It often takes the place of common fastening devices such as nuts, bolts, screws, and rivets. Welding is used extensively in fabrication work. *Fabrication* is the construction of an assembly by fastening separate units together. This is often done to produce a structure that normally would need to be cast. Fabricating is a less expensive means of construction.

WELDING JOINTS

The relative position of the parts being welded determines the type of *welding joint* formed. There are five basic types of welded joints, Figure 29.1.

- Butt joint
- Corner joint
- Tee joint
- Lap joint
- Edge joint

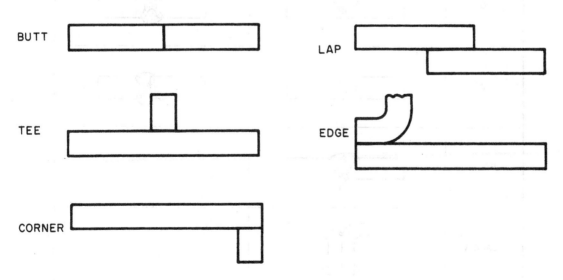

FIGURE 29.1 Types of joints.

TYPES OF WELDS

There are a variety of types of welds that may appear on a print. The selection of a particular weld depends on the joint, material thickness, strength desired, or required penetration. The physical shape of the welded joint gives each weld its name. Figure 29.2 shows some of the basic welds used to join metals.

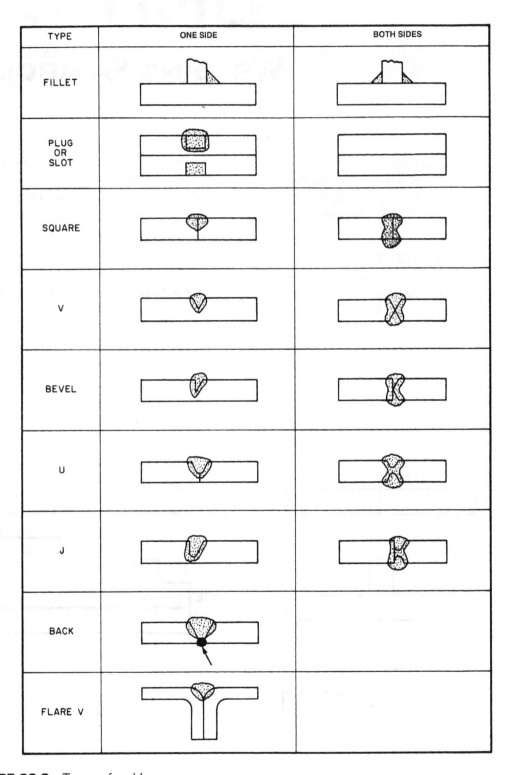

FIGURE 29.2 Types of welds.

WELD AND WELDING SYMBOLS

The standard graphical symbols used to convey welding information were developed by the American National Standards Institute (ANSI Y32.3-1969) and the American Welding Society (AWS A2.4-1979). The symbols are a shorthand method of transmitting information from the drafter to the welder.

The ANSI standard makes a distinction between weld symbols and welding symbols. A **weld symbol** is used to identify the type of weld required. Figure 29.3 shows the basic weld symbols used in industry.

Welding symbols may be made up of several elements of information. The information provides the specific instructions about the type, size, and location of the weld. The elements that may appear on a welding symbol are shown in Figure 29.4.

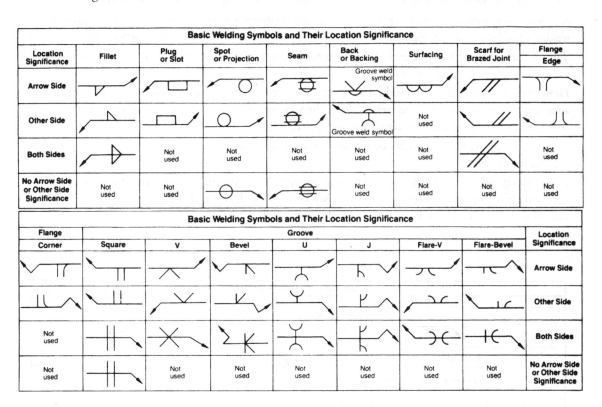

FIGURE 29.3 Basic weld symbols and their location significance.

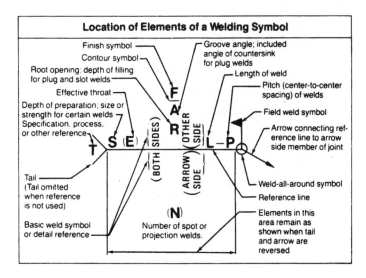

FIGURE 29.4 Location of elements of a welding symbol.

TERMINOLOGY

Reference Line — A heavy solid line that forms the body of the welding symbol. All other information is placed in positions around the reference line.

Arrow — The arrow is attached to the end of the reference line and contacts the weld joint. Welded joints are thus referred to as **arrow side** welds or **other side** welds. The arrow touches the weld joint on the arrow side, as shown in Figure 29.7. The other side weld is located on the part surface opposite the arrow. Arrow side information is always shown below the reference line. Other side information is always shown above the reference line.

Basic Weld Symbols — As previously indicated in Figure 29.3, these specify the type of weld.

Supplementary Symbols — Used to provide additional information as to the extent of welding, place of welding, and bead contour. Figure 29.5 shows the supplementary symbols of the American Welding Society.

Tail — The tail appears on the end of the reference line opposite the arrow. It is used only when a specific welding process is to be specified.

Dimensions — Dimensions of a weld may specify size, length, or spacing of welds. These dimensions appear on the same side of the reference line as the weld symbol. Common practice is to call out the weld size, type, length, and center-to-center spacing (pitch), Figure 29.6.

Finish — Finish requirements may be specified below the arrow side contour symbol or above the other side contour symbol.

Process Specifications — Provided within the tail opening. This information is specified only when necessary. If the welding process is indicated elsewhere on the drawing or the specifications are known, the tail and reference are omitted from the welding symbol.

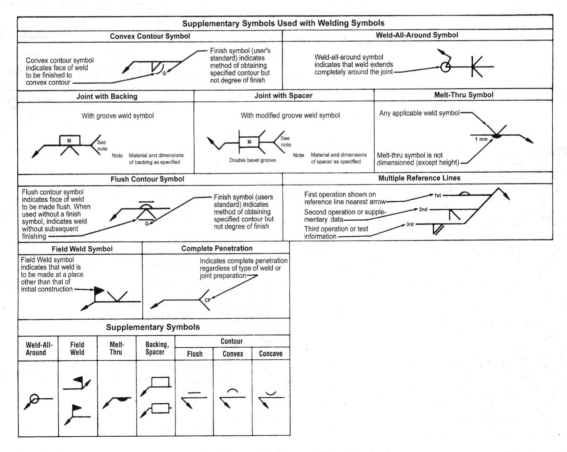

FIGURE 29.5 Supplementary symbols used with welding symbols.

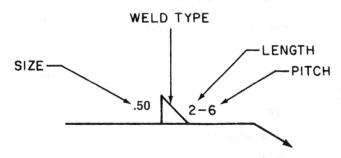

FIGURE 29.6 Dimensioning a fillet weld.

LOCATION OF WELDING SYMBOLS

The welding symbol may be placed on any of the orthographic views. It will generally be shown on the view that best shows the joint. The location of the welding symbol on each view is illustrated in Figure 29.7. However, when it is shown on one view, it is not necessary to include it on any of the other views. In this case, note that the front view is the best view for adding the symbol.

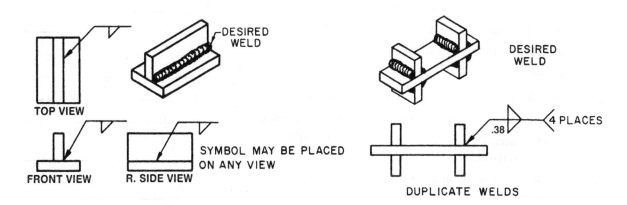

FIGURE 29.7 Location of the weld symbol on orthographic views.

ASSIGNMENT D-24: STOCK PUSHER GUIDE

1. What is the name of the object of which this unit is a part? _____

2. What is the name of this specific assembly? _____

3. What is the name of detail ③? _____

4. How long is detail ④? _____

5. What is the diameter of detail ④? _____

6. How many threads per inch does detail ④ have? _____

7. How is the nut, detail ⑤, secured to detail ③? _____

8. What does the small circle at the joint of the symbol mean? ⌐° _____

9. What kind of weld is called for by the small triangle on the underside of the welding symbol? _____

10. With what type of weld is detail ③ fastened to details ① and ②? _____

11. Is any size given for this weld? _____

12. Is this weld used on one or both sides? _____

13. How are details ①, ②, and ③ fastened together? _____

14. What kind of nut is called for by detail ⑤? _____

15. What are the width, length, and thickness of detail ①? _____

16. What are the width, length, and thickness of detail ②? _____

17. What are the width, length, and depth of detail ③? _____

18. How many tapped holes are specified? _____

19. How deep is the threaded portion of the .26 holes? _____

20. What is the total depth that the tap drill enters the work? _____

21. What is the center-to-center distance of the two tapped holes? _____

22. The back of detail ② is not even, or flush, with detail ①. How much is the offset? _____

23. Detail ③ is mounted at an angle to details ① and ②. At what degree is it mounted? _____

24. What is the distance from the lower left-hand corner of detail ① to the left side of detail ③ in the front view? _____

25. What kind of setscrew is called for by detail ④? _____

26. Is the right end of detail ② even, or flush, with the right end of detail ①? _____

27. What is the distance from the back edge of detail ② to the centerlines of the tapped holes? _____

28. What tolerance is allowed for the center-to-center location of the tapped holes? _____

29. Have any machined surfaces been specified for this part? _____

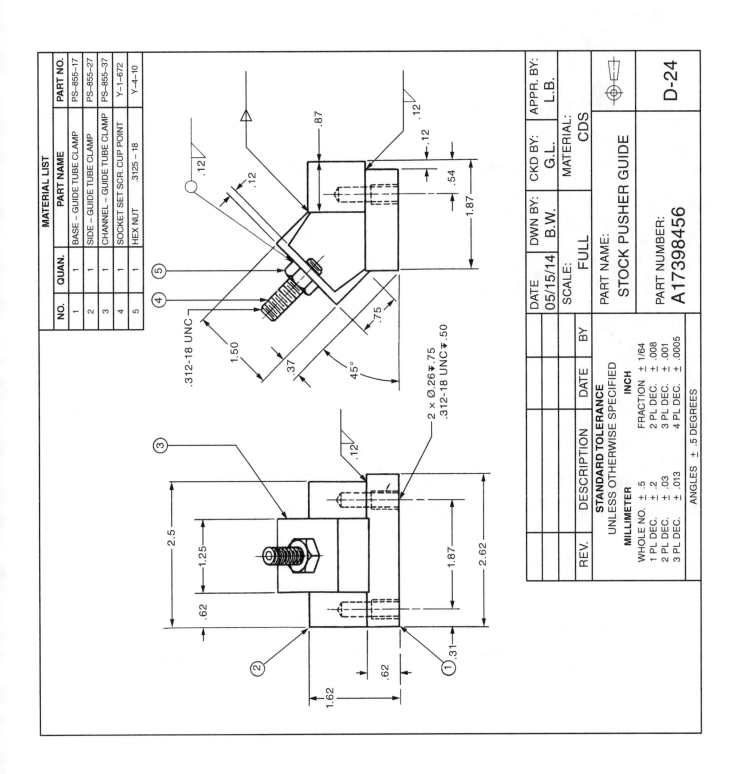

MATERIAL LIST

NO.	QUAN.	PART NAME	PART NO.
1	1	BASE – GUIDE TUBE CLAMP	PS-855-17
2	1	SIDE – GUIDE TUBE CLAMP	PS-855-27
3	1	CHANNEL – GUIDE TUBE CLAMP	PS-855-37
4	1	SOCKET SET SCR. CUP POINT	Y-1-672
5	1	HEX NUT .3125 – 18	Y-4-10

DATE 05/15/14	DWN BY: B.W.	CKD BY: G.L.	APPR. BY: L.B.
SCALE: FULL		MATERIAL: CDS	

PART NAME:
STOCK PUSHER GUIDE

PART NUMBER:
A17398456

D-24

STANDARD TOLERANCE
UNLESS OTHERWISE SPECIFIED

REV.	DESCRIPTION	DATE	BY

MILLIMETER
WHOLE NO. ± .5
1 PL DEC. ± .2
2 PL DEC. ± .03
3 PL DEC. ± .013

INCH
FRACTION ± 1/64
2 PL DEC. ± .008
3 PL DEC. ± .001
4 PL DEC. ± .0005

ANGLES ± .5 DEGREES

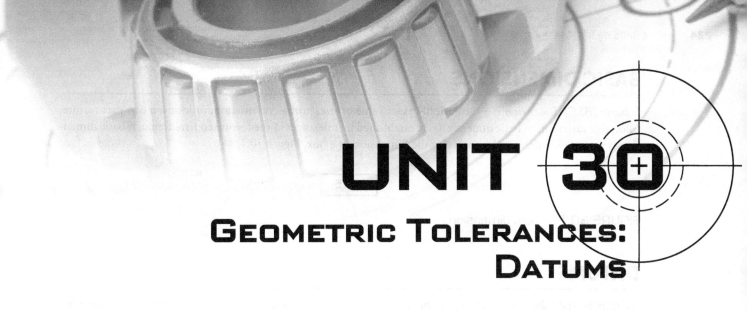

UNIT 30

GEOMETRIC TOLERANCES: DATUMS

Modern-day manufacturing processes require precise tolerances to ensure the interchangeability of parts. Mass-produced parts must be held within specified dimensional tolerances to achieve proper function and relationship to mating units. Geometric dimensioning controls the form or position of part features by means of a language of symbols. These symbols enable the print reader to interpret dimensional requirements and limit the number of notes on a drawing. The geometric system of dimensioning is a widely accepted practice in industry.

This unit describes the key elements that apply to geometric dimensioning.

TERMINOLOGY

Allowance — The intentional difference in size between mating parts.

Basic Dimension — The exact theoretical dimension used to locate a feature or define a true profile.

Nominal Size — The stated designated size of an object, which may or may not be the actual size.

Feature — The specific portion of an object to which dimensions and tolerances are applied. A feature may include one or more surfaces, holes, slots, threads, and so on.

Limits of Size — The applicable maximum and minimum size of a feature.

Form Tolerance — The amount of permissible surface variation from the basic or perfect form.

Positional Tolerance — The amount of permissible dimensional variation from basic or perfect location.

True Position — The term used to describe the perfect location of a point, line, or surface.

Datum — Points, lines, planes, cylinders, axes that are assumed to be exact for purposes of reference. Datums are established from actual features and are used to establish the relationship of other features.

Datum Axis — The theoretically exact centerline of a datum cylinder.

Datum Cylinder (or other geometrical form) — The theoretically exact form profile of the actual datum feature surface.

Datum Feature — The actual part surface or feature used to establish a datum.

Datum Plane — The theoretically exact plane established by the extremities of the actual feature surface.

Specified Datum — A surface or feature identified with a datum symbol A.

Datum Target — Identifies a specific point, line, or area on the object. Datum targets are used to establish a datum for manufacturing and inspection purposes.

Least Material Condition (LMC) — The condition that exists when a part feature contains the minimum amount of material: for example, the maximum diameter of a hole or the minimum diameter of a shaft.

Regardless of Feature Size (RFS) — Specifies that the feature or datum reference applies regardless of where the feature lies within the size tolerance.

Maximum Material Condition (MMC) — The condition that exists when a part feature contains the maximum amount of material: for example, the minimum diameter of a hole or the maximum diameter of a shaft.

BASIC DIMENSIONS

A *basic (BSC) dimension* is a theoretically exact value of size, profile, orientation, or location of a part feature. Allowable variations in basic dimensions are established by tolerances in the feature control frame. A basic dimension is shown on a drawing by enclosing the dimension in a box, Figure 30.1.

$$\boxed{1.375}$$

FIGURE 30.1   Basic dimension.

DATUMS

A datum is established by, or relative to, the actual features of a part. They are used as references from which other features are located. These datums may be points, lines, planes, axes, or cylinders. However, they must not be confused with datum features. A *datum feature* is a real physical part of the object that may have surface variations.

DATUM PLANE

A *datum plane* is an imaginary plane that contacts the datum feature at the highest points of variation, Figure 30.2. One or more datum planes may be used to establish positional relationships on a part. These datum planes are identified as primary, secondary, or tertiary.

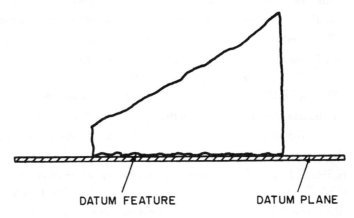

DATUM FEATURE   DATUM PLANE

FIGURE 30.2   Datum plane and datum feature.

   Primary datum planes are developed by establishing three points of contact on the primary datum surface. The contact points must not be in the same line, Figure 30.3.

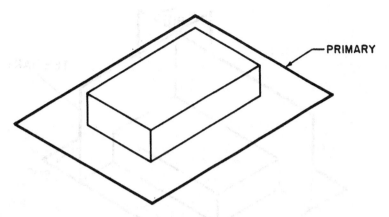

FIGURE 30.3 Primary datum plane.

Secondary datum planes are established on the secondary datum feature. The secondary plane is perpendicular to the primary plane. Two points of contact are used to establish the secondary datum plane, Figure 30.4.

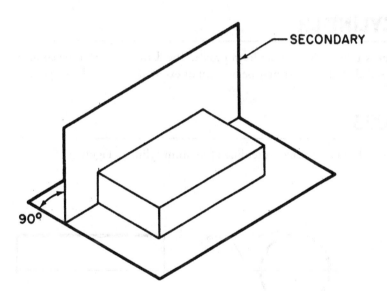

FIGURE 30.4 Secondary datum plane.

Tertiary datum planes are perpendicular to both the primary and secondary datum planes. One point of contact is used to establish the tertiary datum plane, Figure 30.5.

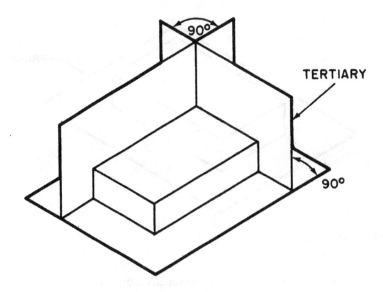

FIGURE 30.5 Tertiary datum plane.

DATUM CYLINDER

A *datum cylinder* is a theoretically exact form profile. The datum is formed by contact with the high points of the datum feature. A datum cylinder may be internal or external as in a hole or cylindrical shaft, Figure 30.6.

DATUM AXIS

A *datum axis* is the theoretical exact center line of a datum cylinder, Figure 30.6.

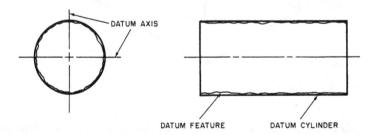

FIGURE 30.6 Datum axis and datum cylinder.

DATUM TARGETS

Datum targets are used to establish the position of a part in a datum reference frame by identifying datum target points, lines, or areas on the part. Specified datum targets are often applied to parts such as castings, forgings, and other parts having irregular contours.

Datum targets ensure repeatability of part location for machining and inspection purposes. A circular area datum target symbol is shown in Figure 30.7.

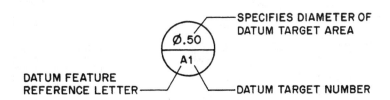

FIGURE 30.7 Datum target.

DATUM IDENTIFICATION SYMBOL

Datum features must be identified on drawings with a datum symbol. These symbols indicate the datum surface being referenced. The symbol used is a capital letter enclosed in a box (frame), Figure 30.8.

A leader line is used to connect the datum frame to the datum feature referenced. A small triangle, attached to the end of the leader line, establishes the connection to the datum feature. The datum symbol may be attached to a surface extension line or directly to the surface being referenced on the object, Figure 30.9.

FIGURE 30.8 Datum identification symbol.

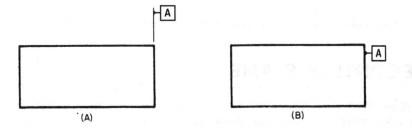

FIGURE 30.9 Placement of datum symbols.

GEOMETRIC CHARACTERISTIC SYMBOLS

Geometric characteristic symbols are used in geometric dimensioning to specify tolerances applied to a part feature. The symbols eliminate the need for written notes on drawings. Tolerances specified may be tolerances of position or form.

FEATURES	TYPES OF TOLERANCE	CHARACTERISTICS	SYMBOLS	SEE UNITS
INDIVIDUAL FEATURES	FORM	STRAIGHTNESS	—	37 & 38
		FLATNESS	▱	39
		CIRCULARITY (ROUNDNESS)	○	39
		CYLINDRICITY	⌭	39
INDIVIDUAL OR RELATED FEATURES	PROFILE	PROFILE OF A LINE	⌒	44
		PROFILE OF A SURFACE	⌓	44
RELATED FEATURES	ORIENTATION	ANGULARITY	∠	41
		PERPENDICULARITY	⊥	
		PARALLELISM	//	
	LOCATION	POSITION	⊕	43
		CONCENTRICITY	◎	
		SYMMETRY	�	
	RUNOUT	CIRCULAR RUNOUT	*↗	45
		TOTAL RUNOUT	*↗↗	
SUPPLEMENTARY SYMBOLS		MAXIMUM MATERIAL CONDITION	Ⓜ	38 & 43
		LEAST MATERIAL CONDITION	Ⓛ	
		PROJECTED TOLERANCE ZONE	Ⓟ	
		BASIC DIMENSION	XX	40
		DATUM FEATURE	▷A	40
		DATUM TARGET	Ø.50 / A2	42

* MAY BE FILLED IN

FIGURE 30.10 Feature control symbols.

The symbols used and the characteristics they control are shown in Figure 30.10.

FEATURE CONTROL FRAME

The method of applying geometric characteristic symbols and tolerances to drawings is the same as that used for datums. geometric characteristic symbols are enclosed in a *frame*. The frame may be divided into two or more separate parts. The first space within the frame shows the geometric symbol. The second space specifies the tolerance applied to the feature. If the tolerance applies to a diameter, the symbol for diameter precedes the tolerance dimension. Figure 30.11 shows typical geometric characteristic symbols.

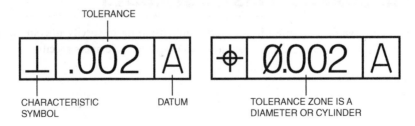

TOLERANCE

⊥ .002 A   ⊕ Ø.002 A

CHARACTERISTIC SYMBOL   DATUM   TOLERANCE ZONE IS A DIAMETER OR CYLINDER

FIGURE 30.11 Typical feature control frames.

Feature control frames may be located on extension or dimension lines applied to part features. They may also be referenced with leader lines or located adjacent to dimensional notes pertaining to the part feature, Figure 30.12.

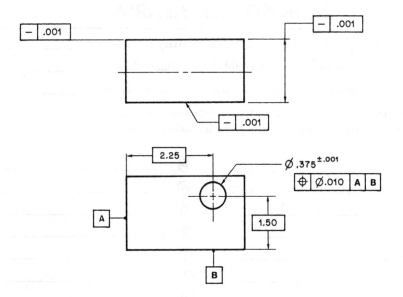

FIGURE 30.12 Placement of feature control frames.

ASSIGNMENT D-25: POSITIONING ARM

1. What surface in the top view is used to establish the tertiary datum? _____

2. What diameter in the top view is used to establish the primary datum? _____

3. What surface in the front view is used to establish the secondary datum? _____

4. Projected surfaces may be identified by determining distances.
   Determine the distances indicated by each of the following letters:

   Ⓐ = _____   Ⓜ = _____

   Ⓑ = _____   Ⓝ = _____

   Ⓒ = _____   Ⓠ = _____

   Ⓓ = _____   Ⓡ = _____

   Ⓔ = _____   Ⓣ = _____

   Ⓕ = _____   Ⓤ = _____

   Ⓖ = _____   Ⓥ = _____

   Ⓗ = _____   Ⓦ = _____

   Ⓘ = _____   Ⓧ = _____

   Ⓙ = _____   Ⓨ = _____

   Ⓚ = _____   Ⓩ = _____

   Ⓛ = _____

5. What is the nominal size of the hole used as datum feature [A]? _____

6. What is the maximum material condition (MMC) of the two .38 holes? _____

7. What surface in the top view is used to establish datum feature [C]? _____

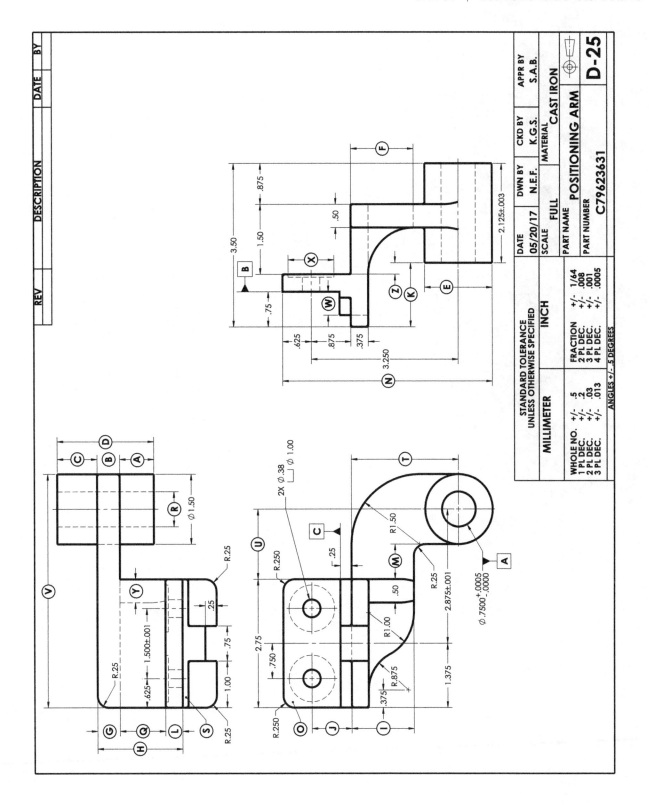

STANDARD TOLERANCE UNLESS OTHERWISE SPECIFIED			
MILLIMETER		INCH	
WHOLE NO.	+/- .5	FRACTION	+/- 1/64
1 PL DEC.	+/- .2	2 PL DEC.	+/- .008
2 PL DEC.	+/- .03	3 PL DEC.	+/- .001
3 PL DEC.	+/- .013	4 PL DEC.	+/- .0005
		ANGLES +/- .5 DEGREES	

DATE 05/20/17	DWN BY N.E.F.	CKD BY K.G.S.	APPR BY S.A.B.
SCALE FULL		MATERIAL CAST IRON	
PART NAME		POSITIONING ARM	
PART NUMBER		C79623631	

D-25

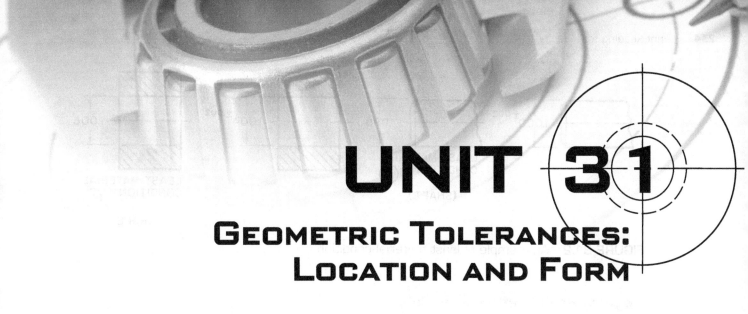

UNIT 31

GEOMETRIC TOLERANCES: LOCATION AND FORM

The previous unit discussed the basic symbols and terminology used in geometric dimensioning. This unit provides a greater understanding of each characteristic and how tolerances are identified.

MODIFIERS

Symbols called *modifiers* are used to indicate that tolerance requirements apply when a part feature is at a specific condition of size.

Maximum Material Condition (MMC)

The *maximum material condition* of a part exists when a feature contains the maximum material allowed. An example is a pin or shaft at its high limit dimension, or a slot or hole at its lowest limit, Figure 31.1. Maximum material condition is specified by the modifier symbol Ⓜ. It is also abbreviated with the letters MMC.

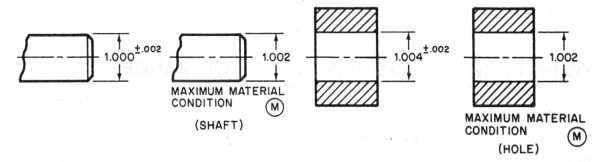

FIGURE 31.1   An example of maximum material condition.

The MMC applies when:

1. Two or more features are interrelated with respect to location or form. For example, a hole and an edge or two holes, and so on. At least one of the related features is to be one of size.
2. The feature to which the MMC applies must be a feature of size. For example, a hole, slot, or pin with an axis.

Least Material Condition (LMC)

3. The *least material condition* of a part exists when a feature contains the minimum material allowed. An example would be a pin or shaft at its low limit dimension or a slot or hole at its highest limit, Figure 31.2. Least material condition is specified by the modifier symbol Ⓛ. It is also abbreviated as LMC.

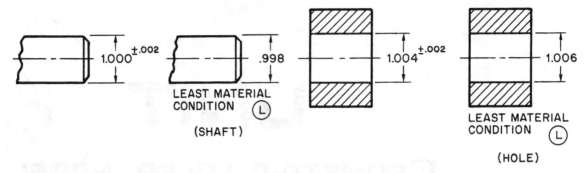

FIGURE 31.2   An example of least material condition.

Regardless of Feature Size (RFS)

4. The *regardless of feature size* symbol is no longer required on industrial drawings. However, it may still be found on various prints or may be used by some companies. RFS is a condition where the tolerance of form or position must be met regardless of where the feature lies within the size tolerance. The modifier symbol for RFS is Ⓢ.

When modifiers are specified on a drawing, they appear in the same box and to the right of the tolerance, Figure 31.3.

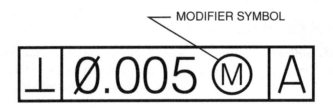

FIGURE 31.3   Specifying modifiers on a drawing.

FORM, PROFILE, AND ORIENTATION TOLERANCES

Tolerances of *form*, *profile*, and *orientation* specify the allowable variation in the geometric shape of the part feature.

Form Tolerances

Form tolerances specify the allowable variation from perfect form as shown on the print, Figure 31.4. Form tolerances are critical to part interchangeability.

Tolerances of form include flatness, straightness, roundness or circularity, and cylindricity.

Flatness. ▱   *Flatness* is the condition of having all elements of a surface in one plane, Figure 31.4A. The tolerance for flatness is defined as the dimensional area formed by two flat planes. The entire surface of the feature, including variations, must fall in this zone.

Straightness. ——————   Straightness is different than flatness and should not be confused. *Straightness* refers to an element of a surface being in a straight line, Figure 31.4B. The tolerance for straightness specifies a zone of uniform width along the length of the feature. All points of the feature as measured along that line must fall within that zone.

Roundness or Circularity. ○ *Roundness* is the condition where each circular element of the surface is an equal distance from the center, Figure 31.4C. The tolerance zone is formed by two concentric circles. The actual surface elements must lie between these two circles at any place of cross section.

Cylindricity. /○/ *Cylindricity* is the condition where all elements of a surface of revolution form a cylinder. The tolerance zone is defined by two concentric cylinders along the length of the feature, Figure 31.4D. All points of the part feature must fall between these cylinders.

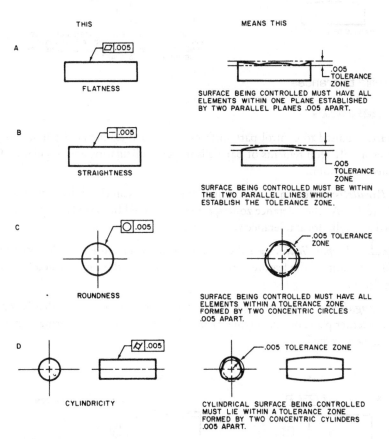

FIGURE 31.4 Form tolerances.

Profile Tolerances

Profile tolerances are used to control the shape of arcs, part contour, or other irregular surfaces. Geometric tolerance symbols may specify either profile control of a line or profile control of an entire surface.

Profile of a Line. ⌒ The *profile of a line* is the feature profile as measured along a line, Figure 31.5A. The tolerance of the profile is the variation from perfect form. All points along the part feature must fall between parallel lines of the perfect profile.

Profile of a Surface. ⌓ The *profile of a surface* is much like the profile of a line. However, the definition is broadened to cover the entire feature surface, Figure 31.5B. All points of the feature surface must fall within the tolerance zone of perfect profile.

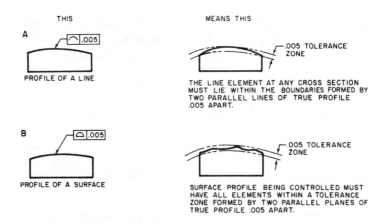

FIGURE 31.5 Profile tolerances.

Orientation Tolerances

Orientation tolerances are used to control part surfaces, individual elements of parts, and part size features. Orientation tolerances specify requirements of parallelism, perpendicularity, and angularity. Orientation tolerances are always related to a datum.

Parallelism. // *Parallelism* refers to a surface, line, or axis that is an equal distance from a datum plane or axis at all points, Figure 31.6A. The tolerance zone specified is defined by two planes parallel to a datum plane. It may also be defined as a cylindrical tolerance zone parallel to a datum axis.

Perpendicularity. ⊥ *Perpendicularity* is the condition where a feature is 90 degrees from a datum plane or axis, Figure 31.6B. All points of the feature surface must fall within the zone formed by two planes perpendicular to the datum.

Angularity. ∠ *Angularity* is the condition of a surface, axis, or center plane that is at an angle other than 90 degrees from a center plane or axis, Figure 31.6C. The tolerance zone is formed by two parallel lines inclined at the exact angle specified.

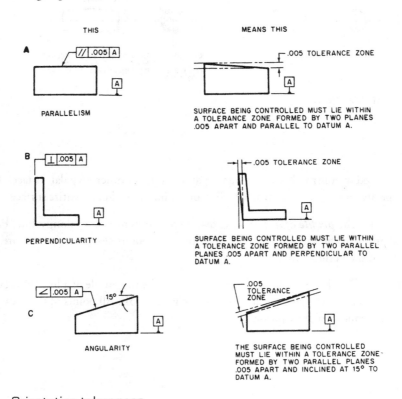

FIGURE 31.6 Orientation tolerances.

RUNOUT TOLERANCES

Runout tolerances specify the allowable variation from perfect form or orientation as described on the print. The two runout tolerance symbols used are *total runout* and *circular runout*. Runout tolerances are always related to a datum.

Circular Runout. (* ⤴) is the deviation of the part feature at any measuring position when rotated 360 degrees on a datum axis, Figure 31.7A. A dial indicator is used to read runout on a feature. The tolerance zone is formed by two coaxial circles within which the total feature runout must fall.

Total Runout. (* ⤴⤴) is the deviation of the entire surface at any measuring point within the specified tolerance zone made up of two coaxial cylinders when rotated 360 degrees on a datum axis, Figure 31.7B.

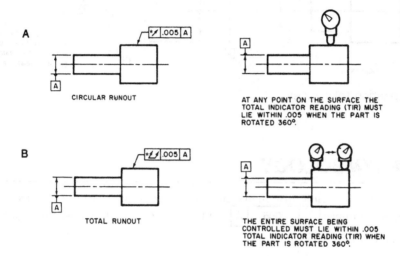

FIGURE 31.7 Runout tolerances.

LOCATION TOLERANCES

Location tolerances specify the allowable variation in the location of a part feature in relation to another feature or datum. The two geometric tolerance symbols used to specify location are *position* and *concentricity*. Location tolerances must always involve at least one size feature and frequently apply an MMC modifier to maximize the interchangeability of parts.

Position. ⌖   *True position* is the term applied to the exact or perfect location of a feature. The true position is located with reference to one or more datums. The position tolerance is the maximum amount of variation allowed. From true position, the amount of permissible tolerance is called a *tolerance zone*. For cylindrical features, this zone is a diameter within which the axis of the feature must lie, Figure 31.8A.

Concentricity. ◎   *Concentricity* refers to the condition of two features sharing a common axis. An example would be a stepped shaft where two diameters share a common centerline, Figure 31.8B. The concentricity tolerance is the diameter of the concentricity tolerance zone within which the feature axis must be.

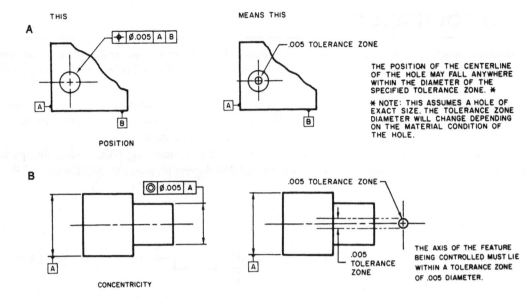

FIGURE 31.8 Location tolerances.

REVIEW OF SYMBOLOGY

In the following questions, use the information in each statement to create the correct datums (example: [A]) and feature control symbols (example: [⊥ | .005 | A]) and add these to the accompanying figures.

1. Make the top flat to within .005.

2. Make the end perpendicular to the bottom within .005.

3. Make the periphery of this cylinder round to within .005.

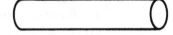

4. Make the top and bottom surfaces flat and equal distance from each other within .005.

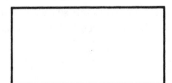

5. Make the 30-degree angle correct to within .005 of the bottom surface.

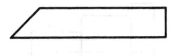

6. Make smaller diameter concentric to the larger diameter within .005.

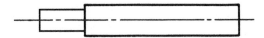

7. Make this shaft cylindrical within .005.

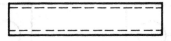

8. Make the two top surfaces parallel to the bottom within .002.

9. Make the profile of the top surface within .004 of the end.

10. Make the larger diameter of the shaft be within .002 runout with the smaller diameter.

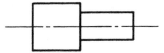

11. Make the top parallel, the right end perpendicular, and left angular to the bottom datum within .003.

12. Make the shaft straight within .003.

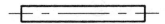

13. Make the top parallel to the bottom within .006.

14. Make runout on the faces and diameters within .002 of the bore.

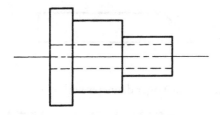

15. Establish two surfaces as datum features. Locate the holes and diametral positional tolerances of .002 at MMC.

ASSIGNMENT D-26: TRIP BOX

1. What line in the top view represents surface Ⓓ? _____

2. Locate surface Ⓐ in the left and front views. _____

3. Locate surface Ⓑ in the front view. _____

4. What line in the left view represents surface ③? _____

5. What is the center distance between holes Ⓑ and Ⓓ? _____

6. Determine distance ④. _____

7. Determine distance ⑤. _____

8. Determine distance ⑥. _____

9. Determine distance ⑪. _____

10. What surface of the left view does line ⑭ represent? _____

11. What point in the front view represents line ⑮? _____

12. What is the thickness of boss Ⓔ? _____

13. Locate surface Ⓖ in the left view. _____

14. Determine distance Ⓜ. _____

15. What point or line in the top view does point ⑯ represent? _____

16. What tolerance of parallelism is allowed on the surface of boss Ⓔ? _____

17. What surface in the left view is used as the primary datum? _____

18. What does the geometric symbol Ⓜ indicate? _____

19. What surface finish is required on the surface shown by hidden line ②? _____

20. What diameter is allowed on the Ø.750 hole at maximum material condition (MMC)? _____

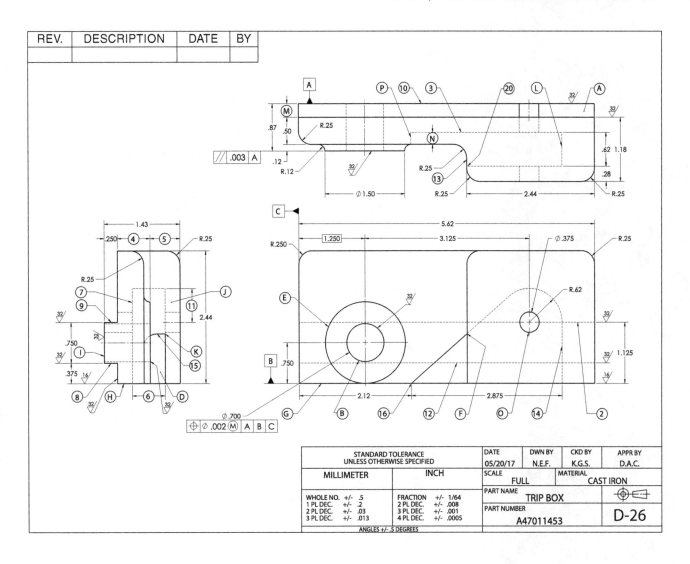

REV.	DESCRIPTION	DATE	BY

STANDARD TOLERANCE UNLESS OTHERWISE SPECIFIED		DATE 05/20/17	DWN BY N.E.F.	CKD BY K.G.S.	APPR BY D.A.C.
MILLIMETER	**INCH**	SCALE FULL		MATERIAL CAST IRON	
WHOLE NO. +/- .5 1 PL DEC. +/- .2 2 PL DEC. +/- .03 3 PL DEC. +/- .013	FRACTION +/- 1/64 2 PL DEC. +/- .008 3 PL DEC. +/- .001 4 PL DEC. +/- .0005	PART NAME TRIP BOX			
		PART NUMBER A47011453		D-26	
ANGLES +/- .5 DEGREES					

APPENDIX

Table 1   Converting Inch Dimensions to Millimeters

Fraction	Decimals Inch		Millimeters	Fraction	Decimals Inch		Millimeters
	Two Place	Three Place			Two Place	Three Place	
1/64	.02	.016	.04	33/64	.52	.516	13.1
1/32	.03	.031	.08	17/32	.53	.531	13.5
3/64	.05	.047	1.2	35/64	.55	.547	13.9
1/16	.06	.062	1.6	9/16	.56	.562	14.3
5/64	.08	.078	2	37/64	.58	.578	14.7
3/32	.09	.094	2.4	19/32	.59	.594	15.1
7/64	.11	.109	2.8	39/64	.61	.609	15.5
1/8	.12	.125	3.2	5/8	.62	.625	15.9
9/64	.14	.141	3.6	41/64	.64	.641	16.3
5/32	.16	.156	4	21/32	.66	.656	16.7
11/64	.17	.172	4.4	43/64	.67	.672	17.1
3/16	.19	.188	4.8	11/16	.69	.688	17.5
13/64	.20	.203	5.2	45/64	.70	.703	17.9
7/32	.22	.219	5.6	23/32	.72	.719	18.3
15/64	.23	.234	6	47/64	.73	.734	18.7
1/4	.25	.250	6.4	3/4	.75	.750	19.1
17/64	.27	.266	6.8	49/64	.77	.766	19.5
9/32	.28	.281	7.1	25/32	.78	.781	19.9
19/64	.30	.297	7.5	51/64	.80	.797	20.2
5/16	.31	.312	7.9	13/16	.81	.812	20.6
21/64	.33	.328	8.3	53/64	.83	.828	21
11/32	.34	.344	8.7	27/32	.84	.844	21.4
23/64	.36	.359	9.1	55/64	.86	.859	21.8
3/8	.38	.375	9.5	7/8	.88	.875	22.2
25/64	.39	.391	9.9	57/64	.89	.891	22.6
13/32	.41	.406	10.3	29/32	.91	.906	23
27/64	.42	.422	10.7	59/64	.92	.922	23.4
7/16	.44	.438	11.1	15/16	.94	.938	23.8
29/64	.45	.453	11.5	61/64	.95	.953	24.2
15/32	.47	.469	11.9	31/32	.97	.969	24.6
31/64	.48	.484	12.3	63/64	.98	.984	25
1/2	.50	.500	12.7	1	1.00	1.000	25.4

Table 2 Abbreviations and Symbols

And . . . .&	Long . . . . .LG
Across Flats . . . . .ACR FLT	Machined . . . .√
Approximate . . . . .APPROX	Machine Steel . . . . .MST
Assembly . . . . .ASSY	Malleable Iron . . . . .MI
Bill of Material . . . . .B/M	Material . . . . .MATL
Bolt Circle . . . . .BC	Maximum . . . . .MAX
Brass . . . . .BR	Maximum Material Condition . . .MMC or Ⓜ
Bronze . . . . .BRZ	Meter . . . . .m
Brown and Sharpe Gage . . . . . . .B & S GA	Metric Thread . . . . .M
Carbon Steel . . . . .CS	Micrometer . . . . .μm
Cast Iron CI	Mild Steel . . . . .MS
Centimeter . . . . .cm	Millimeter . . . . .mm
Center Line . . . .₵ or CL	Minimum . . . . .MIN
Center to Center . . . . .C to C	Minute (Angle) . . . .(′)
Chamfer . . . . .CHAM	Nominal . . . . .NOM
Circularity . . . . .CIR	Not to Scale XXX
Cold Rolled Steel . . . . .CRS	Number . . . . .NO
Concentric . . . . .CONC	Outside Diameter . . . . .OD
Conical Taper . . . .▷	Parallel .PAR
Copper .COP	Perpendicular . . . . .PERP
Counterbore . . . . .CBORE or ⌴	Pitch Circle Diameter . . . . .PCD
Countersink . . . . .CSK or ∨	Pitch Diameter . . . . .PD
Cubic Centimeter . . . . .cm^3	Projected Tolerance Zone . . . .Ⓟ
Cubic Meter . . . . .m^3	Radius .R
Datum . . . . A	Referencde or Reference
Deep or Depth . . . .▽ △	Dimension . . . .( )
Degree (Angle) . . . .°	Regardless of Feature Size . . . . . .RFS
Diameter . . . .∅ or DIA	Revolutions per Minute . . . . .R/MIN
Diametral Pitch . . . . .DP	Right Hand . . . . .RH
Dimension . . . . .DIM	Second (Arc) . . . .″
Dimension Origin . . . .⟜	Section .SECT
Drawing . . . . .DWG	Slope . . . .◁
Eccentric . . . . .ECC	Spherical Radius . . . . .SR
Equally Spaced . . . . .EQL SP	Spotface . . . . .SF or ⌴
Figure .FIG	Square .SQ or □
Finish All Over . . . . .FAO	Square Centimeter . . . . .cm^2
Gage . . . . .GA	Square Meter . . . . .m^2
Gray Iron . . . . .GI	Steel . . . . .STL
Head . . . . .HD	Symmetrical . . . . .SYM or ⌗
Heat Treat . . . . .HT TR	Symmetry . . . .⌗
Hexagon . . . . .HEX	Taper Pipe Thread . . . . .NPT
Inside Diameter . . . . .ID	Thick . . . . .THK
International Organization for	Through . . . . .THRU
Standardization . . . . .ISO	Undercut . . . . .UCUT
International Pipe Standard . . . . .IPS	United States Gage . . . . .USG
Kilogram . . . . .kg	Wrought Iron . . . . .WI
Kilometer . . . . .km	Wrought Steel . . . . .WS
Least Material Condition . . . . . . .LMC or Ⓛ	
Left Hand . . . . .LH	

Source: American Society of Mechanical Engineers

Table 3   Number and Letter-Size Drills

	Decimal Inch and Millimeter Equivalents of Number Size Drills					Decimal Inch and Millimeter Equivalents of Letter Size Drills		
No.	Decimal Inch	mm	No.	Decimal Inch	mm	Letter	Decimal Inch	mm
1	.2280	5.8	31	.1200	3.0	A	.234	5.9
2	.2210	5.6	32	.1160	2.9	B	.238	6.0
3	.2130	5.4	33	.1130	2.9	C	.242	6.1
4	.2090	5.3	34	.1110	2.8	D	.246	6.2
5	.2055	5.2	35	.1100	2.8	E	.250	6.4
6	.2040	5.2	36	.1065	2.7	F	.257	6.5
7	.2010	5.1	37	.1040	2.6	G	.261	6.6
8	.1990	5.1	38	.1015	2.6	H	.266	6.8
9	.1960	5.0	39	.0095	2.5	I	.272	6.9
10	.1935	4.9	40	.0980	2.5	J	.277	7.0
11	.1910	4.9	41	.0960	2.4	K	.281	7.1
12	.1890	4.8	42	.0935	2.4	L	.290	7.4
13	.1850	4.7	43	.0890	2.3	M	.295	7.5
14	.1820	4.6	44	.0860	2.2	N	.302	7.7
15	.1800	4.6	45	.0820	2.1	O	.316	8.0
16	.1770	4.5	46	.0810	2.1	P	.323	8.2
17	.1730	4.4	47	.0785	2.0	Q	.332	8.4
18	.1695	4.3	48	.0760	1.9	R	.339	8.6
19	.1660	4.2	49	.0730	1.9	S	.348	8.8
20	.1610	4.1	50	.0700	1.8	T	.358	9.1
21	.1590	4.0	51	.0670	1.7	U	.368	9.3
22	.1570	4.0	52	.0635	1.6	V	.377	9.6
23	.1540	3.9	53	.0595	1.5	W	.386	9.8
24	.1520	3.9	54	.0550	1.4	X	.397	10.1
25	.1495	3.8	55	.0520	1.3	Y	.404	10.3
26	.1470	3.7	56	.0465	1.2	Z	.413	10.5
27	.1440	3.7	57	.0430	1.1			
28	.1405	3.6	58	.0420	1.1			
29	.1360	3.5	59	.0410	1.0			
30	.1285	3.3	60	.0400	1.0			

Source: American Society of Mechanical Engineers

Table 4 Metric Twist Drill Sizes

Metric Drill Sizes (mm)		Reference Decimal Equivalent (Inches)	Metric Drill Sizes (mm)		Reference Decimal Equivalent (Inches)	Metric Drill Sizes (mm)		Reference Decimal Equivalent (Inches)
Preferred	Available		Preferred	Available		Preferred	Available	
–	0.40	.0157	2.2	–	.0866	10	–	.3937
–	0.42	.0165	–	2.3	.0906	–	10.3	.4055
–	0.45	.0177	2.4	–	.0945	10.5	–	.4134
–	0.48	.0189	2.5	–	.0984	–	10.8	.4252
0.5	–	.0197	2.6	–	.1024	11	–	.4331
–	0.52	.0205	–	2.7	.1063	–	11.5	.4528
0.55	–	.0217	2.8	–	.1102	12	–	.4724
–	0.58	.0228	–	2.9	.1142	12.5	–	.4921
0.6	–	.0236	3	–	.1181	13	–	.5118
–	0.62	.0244	–	3.1	.1220	–	13.5	.5315
0.65	–	.0256	3.2	–	.1260	14	–	.5512
–	0.68	.0268	–	3.3	.1299	–	14.5	.5709
0.7	–	.0276	3.4	–	.1339	15	–	.5906
–	0.72	.0283	–	3.5	.1378	–	15.5	.6102
0.75	–	.0295	3.6	–	.1417	16	–	.6299
–	0.78	.0307	–	3.7	.1457	–	16.5	.6496
0.8	–	.0315	3.8	–	.1496	17	–	.6693
–	0.82	.0323	–	3.9	.1535	–	17.5	.6890
0.85	–	.0335	4	–	.1575	18	–	.7087
–	0.88	.0346	–	4.1	.1614	–	18.5	.7283
0.9	–	.0354	4.2	–	.1654	19	–	.7480
–	0.92	.0362	–	4.4	.1732	–	19.5	.7677
0.95	–	.0374	4.5	–	.1772	20	–	.7874
–	0.98	.0386	–	4.6	.1811	–	20.5	.8071
1	–	.0394	4.8	–	.1890	21	–	.8268
–	1.03	.0406	5	–	.1969	–	21.5	.8465
1.05	–	.0413	–	5.2	.2047	22	–	.8661
–	1.08	.0425	5.3	–	.2087	–	23	.9055
1.1	–	.0433	–	5.4	.2126	24	–	.9449
–	1.15	.0453	5.6	–	.2205	25	–	.9843
1.2	–	.0472	–	5.8	.2283	26	–	1.0236
1.25	–	.0492	6	–	.2362	–	27	1.0630
1.3	–	.0512	–	6.2	.2441	28	–	1.1024
–	1.35	.0531	6.3	–	.2480	–	29	1.1417
1.4	–	.0551	–	6.5	.2559	30	–	1.1811
–	1.45	.0571	6.7	–	.2638	–	31	1.2205
1.5	–	.0591	–	6.8	.2677	32	–	1.2598
–	1.55	.0610	–	6.9	.2717	–	33	1.2992
1.6	–	.0630	7.1	–	.2795	34	–	1.3386
–	1.65	.0650	–	7.3	.2874	–	35	1.3780
1.7	–	.0669	7.5	–	.2953	36	–	1.4173
–	1.75	.0689	–	7.8	.3071	–	37	1.4567
1.8	–	.0709	8	–	.3150	38	–	1.4961
–	1.85	.0728	–	8.2	.3228	–	39	1.5354
1.9	–	.0748	8.5	–	.3346	40	–	1.5748
–	1.95	.0768	–	8.8	.3465	–	41	1.6142
2	–	.0787	9	–	.3543	42	–	1.6535
–	2.05	.0807	–	9.2	.3622	–	43.5	1.7126
2.1	–	.0827	9.5	–	.3740	45	–	1.7717
–	2.15	.0846	–	9.8	.3858	–	46.5	1.8307

Source: American Society of Mechanical Engineers

Table 5   Unified and American (Inch) Threads

Size		Coarse Thread Series (UNC & NC)		Fine Thread Series (UNF & NF)		Extra Fine Series (UNEF & NEF)		8-Pitch Thread Series (8 N)		12-Pitch Thread Series (12 N)		16-Pitch Thread Series (16 N)	
Number or Fraction	Decimal	Threads Per Inch	Tap Drill	Threads Per Inch	Tap Drill	Threads Per Inch	Tap Drill	Threads Per Inch	Tap Drill	Threads Per Inch	Tap Drill	Threads Per Inch	Tap Drill
0	.060			80	3/64								
1	.073	64	No. 53	72	No. 53								
2	.086	56	No. 50	64	No. 50								
3	.099	48	No. 47	56	No. 45								
4	.112	40	No. 43	48	No. 42								
5	.125	40	No. 38	44	No. 37								
6	.138	32	No. 36	40	No. 33								
8	.164	32	No. 29	36	No. 29								
10	.190	24	No. 25	32	No. 21								
12	.216	24	No. 16	28	No. 14	32	No. 13						
1/4	.250	20	No. 7	28	No. 3	32	7/32						
5/16	.312	18	F	24	I	32	9/32						
3/8	.375	16	5/16	24	Q	32	11/32						
7/16	.438	14	U	20	25/64	28	13/32						
1/2	.500	13	27/64	20	29/64	28	15/32			12	27/64		
9/16	.562	12	31/64	18	33/64	24	33/64			12	31/64		
5/8	.625	11	17/32	18	37/64	24	37/64			12	35/64		
3/4	.750	10	21/32	16	11/16	20	45/64			12	43/64	16	11/16
7/8	.875	9	49/64	14	13/16	20	53/64			12	51/64	16	13/16
1	1.000	8	7/8	12	59/64	20	61/64	8	7/8	12	59/64	16	15/16
1 1/8	1.125	7	63/64	12	1 3/64	18	1 5/64	8	1	12	1 3/64	16	1 1/16
1 1/4	1.250	7	1 7/64	12	1 11/64	18	1 3/16	8	1 1/8	12	1 11/64	16	1 3/16
1 3/8	1.375	6	1 7/32	12	1 19/64	18	1 5/16	8	1 1/4	12	1 19/64	16	1 5/16
1 1/2	1.500	6	1 11/32	12	1 27/64	18	1 7/16	8	1 3/8	12	1 27/64	16	1 7/16
1 3/4	1.750	5	1 9/16			16	1 11/16	8	1 5/8	12	1 43/64	16	1 11/16
2	2.000	4 1/2	1 25/32			16	1 15/16	8	1 7/8	12	1 51/64	16	1 15/16
2 1/4	2.250	4 1/2	2 1/32					8	2 1/8	12	2 11/64	16	2 3/16
2 1/2	2.500	4	2 1/4					8	2 3/8	12	2 27/64	16	2 7/16
2 3/4	2.750	4	2 1/2					8	2 5/8	12	2 43/64	16	2 11/16
3	3.000	4	2 3/4					8	2 7/8	12	2 59/64	16	2 15/16

Color shows unified thread

Source: American Society of Mechanical Engineers

Table 6　Metric Threads

Nominal Size DIA (mm)	Coarse Thread Pitch	Coarse Tap Drill Size	Fine Thread Pitch	Fine Tap Drill Size	4 Thread Pitch	4 Tap Drill Size	3 Thread Pitch	3 Tap Drill Size	2 Thread Pitch	2 Tap Drill Size	1.5 Thread Pitch	1.5 Tap Drill Size	1.25 Thread Pitch	1.25 Tap Drill Size	1 Thread Pitch	1 Tap Drill Size	0.75 Thread Pitch	0.75 Tap Drill Size	0.5 Thread Pitch	0.5 Tap Drill Size	0.35 Thread Pitch	0.35 Tap Drill Size
Series with Graded Pitches	Coarse		Fine		*Series with Constant Pitches*																	
1.6	0.35	1.25																				
1.8	0.35	1.45																				
2	0.4	1.6																				
2.2	0.45	1.75																				
2.5	0.45	2.05																			0.35	2.15
3	0.5	2.5																			0.35	2.65
3.5	0.6	2.9																			0.35	3.15
4	0.7	3.3																	0.5	3.5		
4.5	0.75	3.7																	0.5	4		
5	0.8	4.2																	0.5	4.5		
*6	1	5															0.75	5.2				
**6.3	1	5.3																				
8	1.25	6.7	1	7											1	7	0.75	7.2				
10	1.5	8.5	1.25	8.7									1.25	8.7	1	9	0.75	9.2				
12	1.75	10.2	1.25	10.8							1.5	10.5	1.25	10.7	1	11						
14	2	12	1.5	12.5							1.5	12.5	1.25	12.7	1	13						
16	2	14	1.5	14.5							1.5	14.5			1	15						
18	2.5	15.5	1.5	16.5					2	16	1.5	16.5			1	17						
20	2.5	17.5	1.5	18.5					2	18	1.5	18.5			1	19						
22	2.5	19.5	1.5	20.5					2	20	1.5	20.5			1	21						
24	3	21	2	22					2	22	1.5	22.5			1	23						
27	3	24	2	25					2	25	1.5	25.5			1	26						
30	3.5	26.5	2	28					2	28	1.5	28.5			1	29						
33	3.5	29.5	2	31					2	31	1.5	31.5										
36	4	32	3	33					2	34	1.5	34.5										
39	4	35	3	36					2	37	1.5	37.5										
42	4.5	37.5	3	39	4	38	3	39	2	40	1.5	40.5										
45	4.5	39	3	42	4	41	3	42	2	43	1.5	43.5										
48	5	43	3	45	4	44	3	45	2	46	1.5	46.5										

* ISO thread size

** ASME thread size (to be discontinued)

Source: American Society of Mechanical Engineers

Table 7 Common Cap Screw Sizes

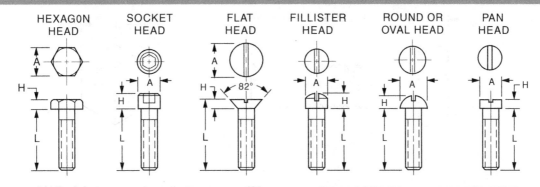

		Hexagon Head		Socket Head		Flat Head		Fillister Head		Round or Oval Head	
Nominal Size											
Fraction	Decimal	A	H	A	H	A	H	A	H	A	H
1/4	.250	.44	.17	.38	.25	.50	.14	.38	.22	.44	.19
5/16	.312	.50	.22	.47	.31	.62	.18	.44	.25	.56	.25
3/8	.375	.56	.25	.56	.38	.75	.21	.56	.31	.62	.27
7/16	.438	.62	.30	.66	.44	.81	.21	.62	.36	.75	.33
1/2	.500	.75	.34	.75	.50	.88	.21	.75	.41	.81	.35
5/8	.625	.94	.42	.94	.62	1.12	.28	.88	.50	1.00	.44
3/4	.750	1.12	.50	1.12	.75	1.38	.35	1.00	.59	1.25	.55
7/8	.875	1.31	.58	1.31	.88	1.62	.42	1.12	.69		
1	1.000	1.50	.67	1.50	1.00	1.88	.49	1.31	.78		
1 1/8	1.125	1.69	.75	1.69	1.12	2.06	.53				
1 1/4	1.250	1.88	.84	1.88	1.25	2.31	.60				
1 1/2	1.500	2.25	1.00	2.25	1.50	2.81	.74				

U.S. Customary (Inches)

	Hexagon Head		Socket Head			Flat Head		Fillister Head		Pan Head	
Nominal Size	A	H	A	H	Key Size	A	H	A	H	A	H
M3	5.5	2	5.5	3	2.5	5.6	1.6	6	2.4	5.6	1.9
4	7	2.8	7	4	3	7.5	2.2	8	3.1	7.5	2.5
5	8.5	3.5	9	5	4	9.2	2.5	10	3.8	9.2	3.1
6	10	4	10	6	5	11	3	12	4.6	11	3.8
8	13	5.5	13	8	6	14.5	4	16	6	14.5	5
10	17	7	16	10	8	18	5	20	7.5	18	6.2
12	19	8	18	12	10						
14	22	9	22	14	12						
16	24	10	24	16	14						
18	27	12	27	18	14						
20	30	13	30	20	17						
22	36	15	33	22	17						
24	36	15	36	24	19						
27	41	17	40	27	19						
30	46	19	45	30	22						

Metric (Millimeters) Sizes

NOTE: Length sizes normally available in .25 inch and 10mm increments

Source: American Society of Mechanical Engineers

Table 8 Hexagon-Head Bolts and Cap Screws

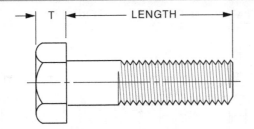
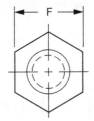

U.S. Customary (Inches)				Metric (Millimeters)		
Nominal Size Fraction	Decimal	Width Across Flats	Thickness	Nominal Size (Millimeters)	Width Across Flats	Thickness
1/4	.250	.44	.17	4	7	2.8
5/16	.312	.50	.22	5	8	3.5
3/8	.375	.56	.25	6	10	4
7/16	.438	.62	.30	8	13	5.5
1/2	.500	.75	.34	10	17	7
5/8	.625	.94	.42	12	19	8
3/4	.750	1.12	.50	14	22	9
7/8	.875	1.31	.58	16	24	10
1	1.000	1.50	.67	18	27	12
1 1/8	1.125	1.69	.75	20	30	13
1 1/4	1.250	1.88	.84	22	32	14
1 3/8	1.375	2.06	.91	24	36	15
1 1/2	1.500	2.25	1.00	27	41	17
				30	46	19
				33	50	21
				36	55	23

NOTE: For bold and cap screw sizes below 7/16 inch and 8mm length sizes normally available in .25 inch and 10mm increments

Source: American Society of Mechanical Engineers

Table 9 Set Screws

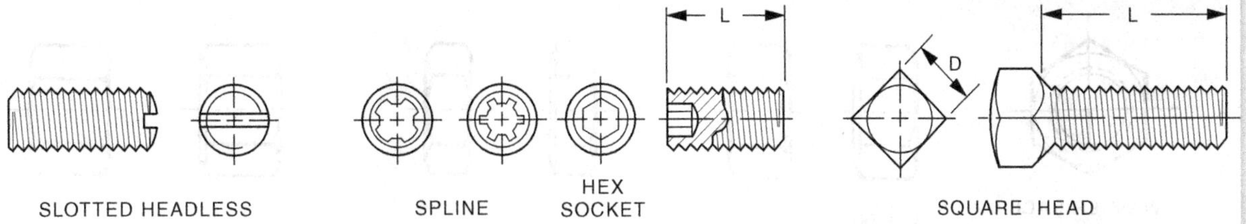

SLOTTED HEADLESS   SPLINE   HEX SOCKET   SQUARE HEAD

SET SCREW HEADS

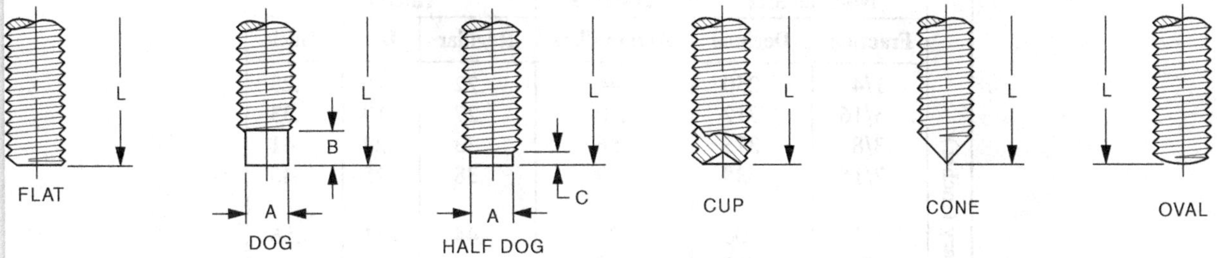

FLAT   DOG   HALF DOG   CUP   CONE   OVAL

SET SCREW POINTS

U.S. Customary (Inches)			Metric (Millimeters)	
Nominal Size		Key Size	Nominal Size	Key Size
Number	Decimal			
4	.112	.050	M 1.4	0.7
5	.125	.062	2	0.9
6	.138	.062	3	1.5
8	.164	.078	4	2
10	.190	.094	5	2.5
12	.216	.109	6	3
1/4	.250	.125	8	4
5/16	.312	.156	10	5
3/8	.375	.188	12	6
1/2	.500	.250	16	8

Source: American Society of Mechanical Engineers

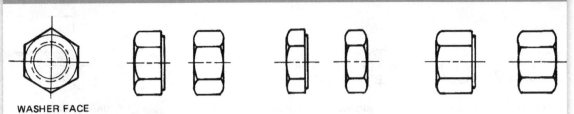

Table 10 Hexagon-Head Nuts

WASHER FACE

REGULAR   JAM   THICK

Nominal Size		Distance Across Flats	Thickness		
Fraction	Decimal		Regular	Jam	Thick
1/4	.250	.44	.22	.16	.28
5/16	.312	.50	.27	.19	.33
3/8	.375	.56	.33	.22	.41
7/16	.438	.69	.38	.25	.45
1/2	.500	.75	.44	.31	.56
9/16	.562	.88	.48	.31	.61
5/8	.625	.94	.55	.38	.72
3/4	.750	1.12	.64	.42	.81
7/8	.875	1.31	.75	.48	.91
1	1.000	1.50	.86	.55	1.00
1 1/8	1.125	1.69	.97	.61	1.16
1 1/4	1.250	1.88	1.06	.72	1.25
1 3/8	1.375	2.06	1.17	.78	1.38
1 1/2	1.500	2.25	1.28	.84	1.50

U.S. Customary (Inches)

Nominal Size (Millimeters)	Distance Across Flats	Thickness		
		Regular	Jam	Thick
4	7	3	2	5
5	8	4	2.5	5
6	10	5	3	6
8	13	6.5	5	8
10	17	8	6	10
12	19	10	7	12
14	22	11	8	14
16	24	13	8	16
18	27	15	9	18.5
20	30	16	9	20
22	32	18	10	22
24	36	19	10	24
27	41	22	12	27
30	46	24	12	30
33	50	26		
36	55	29		
39	60	31		

Metric (Millimeters)

Source: American Society of Mechanical Engineers

Table 11   Hexagon Flanged Nuts

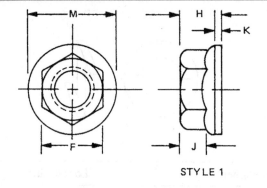

STYLE 1

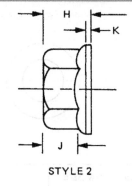

STYLE 2

Metric (Millimeters)							
Nominal Nut Size and Thread Pitch	Width Across Flats F	Style 1				Style 2	
		H	J	K	M	H	J
M6 X 1	10	5.8	3	1	14.2	6.7	3.7
M8 X 1.25	13	6.8	3.7	1.3	17.6	8	4.5
M10 X 1.5	15	9.6	5.5	1.5	21.5	11.2	6.7
M12 X 1.75	18	11.6	6.7	2	25.6	13.5	8.2
M14 X 2	21	13.4	7.8	2.3	29.6	15.7	9.6
M16 X 2	24	15.9	9.5	2.5	34.2	18.4	11.7
M20 X 2.5	30	19.2	11.1	2.8	42.3	22	12.6

Source: American Society of Mechanical Engineers

Table 12   Common Washer

FLAT WASHER   LOCKWASHER

Nominal Screw Size		Flat Washer			Lockwasher		
Number or Fraction	Decimal	Inside Dia A	Outside Dia B	Thickness C	Inside Dia A	Outside Dia B	Thickness C
6	.138	.16	.38	.05	.14	.25	.03
8	.164	.19	.44	.05	.17	.29	.04
10	.190	.22	.50	.05	.19	.33	.05
12	.216	.25	.56	.07	.22	.38	.06
1/4	.250 N	.28	.63	.07	.26	.49	.06
1/4	.250 W	.31	.73	.07			
5/16	.312 N	.34	.69	.07	.32	.59	.08
5/16	.312 W	.38	.88	.08			
3/8	.375 N	.41	.81	.07	.38	.68	.09
3/8	.375 W	.44	1.00	.08			
7/16	.438 N	.47	.92	.07	.45	.78	.11
7/16	.438 W	.50	1.25	.08			
1/2	.500 N	.53	1.06	.10	.51	.87	.12
1/2	.500 W	.56	1.38	.11			
5/8	.625 N	.66	1.31	.10	.64	1.08	.16
5/8	.625 W	.69	1.75	.13			
3/4	.750 N	.81	1.47	.13	.76	1.27	.19
3/4	.750 W	.81	2.00	.15			
7/8	.875 N	.94	1.75	.13	.89	1.46	.22
7/8	.875 W	.94	2.25	.17			
1	1.000 N	1.06	2.00	.13	1.02	1.66	.25
1	1.000 W	1.06	2.50	.17			
1 1/8	1.125 N	1.25	2.25	.13	1.14	1.85	.28
1 1/8	1.125 W	1.25	2.75	.17			
1 1/4	1.250 N	1.38	2.50	.17	1.27	2.05	.31
1 1/4	1.250 W	1.38	3.00	.17			
1 3/8	1.375 N	1.50	2.75	.17	1.40	2.24	.34
1 3/8	1.375 W	1.50	3.25	.18			
1 1/2	1.500 N	1.62	3.00	.17	1.53	2.43	.38
1 1/2	1.500 W	1.62	3.50	.18			

N–SAE Sizes (Narrow)
W–Standard Plate (Wide)   **INCH SIZES**

Source: American Society of Mechanical Engineers

Table 12 (cont'd)   Common Washer

FLAT WASHER   LOCKWASHER   SPRING LOCKWASHER

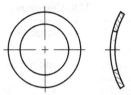

Bolt Size	Flat Washers			Lockwashers			Spring Lockwashers		
	ID	OD	Thickness	ID	OD	Thickness	ID	OD	Thickness
2	2.2	5.5	0.5	2.1	3.3	0.5			
3	3.2	7	0.5	3.1	5.7	0.8			
4	4.3	9	0.8	4.1	7.1	0.9	4.2	8	0.3 0.4
5	5.3	11	1	5.1	8.7	1.2	5.2	10	0.4 0.5
6	6.4	12	1.5	6.1	11.1	1.6	6.2	12.5	0.5 0.7
7	7.4	14	1.5	7.1	12.1	1.6	7.2	14	0.5 0.8
8	8.4	17	2	8.2	14.2	2	8.2	16	0.6 0.9
10	10.5	21	2.5	10.2	17.2	2.2	10.2	20	0.8 1.1
12	13	24	2.5	12.3	20.2	2.5	12.2	25	0.9 1.5
14	15	28	2.5	14.2	23.2	3	14.2	28	1 1.5
16	17	30	3	16.2	26.2	3.5	16.3	31.5	1.2 1.7
18	19	34	3	18.2	28.2	3.5	18.3	35.5	1.2 2
20	21	36	3	20.2	32.2	4	20.4	40	1.5 2.25
22	23	39	4	22.5	34.5	4	22.4	45	1.75 2.5
24	25	44	4	24.5	38.5	5			
27	28	50	4	27.5	41.5	5			
30	31	56	4	30.5	46.5	6			

MILLIMETER SIZES

Source: American Society of Mechanical Engineers

Table 13   Square and Flat Stock Keys

U.S. Customary (Inches)					Metric (Millimeters)					
Diameter of Shaft	Square Key		Flat Key		Diameter of Shaft (mm)		Square Key		Flat Key	
	Nominal Size		Nominal Size				Nominal Size		Nominal Size	
Inclusive	W	H	W	H	Over	Up To	W	H	W	H
.500– .562	.125	.125	.125	.094	12	17	5	5		
.625– .875	.188	.188	.188	.125	17	22	6	6		
.938–1.250	.250	.250	.250	.188	22	30	7	7	8	7
1.312–1.375	.312	.312	.312	.250	30	38	8	8	10	8
1.438–1.750	.375	.375	.375	.250	38	44	9	9	12	8
1.812–2.250	.500	.500	.500	.375	44	50	10	10	14	9
2.312–2.750	.625	.625	.625	.438	50	58	12	12	16	10

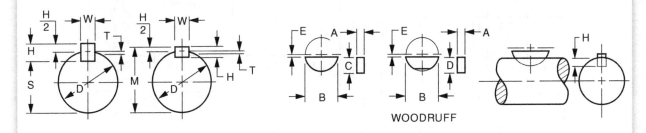

WOODRUFF

C = ALLOWANCE FOR PARALLEL KEYS = .005 IN. OR 0.12 MM

$$S = D - \frac{H}{2} - T = \frac{D - H + \sqrt{D^2 - W^2}}{2} \qquad T = \frac{D - \sqrt{D^2 - W^2}}{2} \qquad \text{W = NORMAL KEY WIDTH (INCH OR MILLIMETERS)}$$

$$M = D - T + \frac{H}{2} + C = \frac{D + H + \sqrt{D^2 - W^2} + C}{2}$$

Source: American Society of Mechanical Engineers

Table 14   Woodruff Keys

Key No.	Nominal (A x B)		U.S. Customary (Inches)				Metric (Millimeters)			
			Key			Keyseat	Key			Key Seat
	Millimeters	Inches	E	C	D	H	E	C	D	H
204	1.6 x 6.4	0.062 x 0.250	.05	.20	.19	.10	0.5	2.8	2.8	4.3
304	2.4 x 12.7	0.094 x 0.500	.05	.20	.19	.15	1.3	5.1	4.8	3.8
305	2.4 x 15.9	0.094 x 0.625	.06	.25	.24	.20	1.5	6.4	6.1	5.1
404	3.2 x 12.7	0.125 x 0.500	.05	.20	.19	.14	1.3	5.1	4.8	3.6
405	3.2 x 15.9	0.125 x 0.625	.06	.25	.24	.18	1.5	6.4	6.1	4.6
406	3.2 x 19.1	0.125 x 0.750	.06	.31	.30	.25	1.5	7.9	7.6	6.4
505	4.0 x 15.9	0.156 x 0.625	.06	.25	.24	.17	1.5	6.4	6.1	4.3
506	4.0 x 19.1	0.156 x 0.750	.06	.31	.30	.23	1.5	7.9	7.6	5.8
507	4.0 x 22.2	0.156 x 0.875	.06	.38	.36	.29	1.5	9.7	9.1	7.4
606	4.8 x 19.1	0.188 x 0.750	.06	.31	.30	.21	1.5	7.9	7.6	5.3
607	4.8 x 22.2	0.188 x 0.875	.06	.38	.36	.28	1.5	9.7	9.1	7.1
608	4.8 x 25.4	0.188 x 1.000	.06	.44	.43	.34	1.5	11.2	10.9	8.6
609	4.8 x 28.6	0.188 x 1.250	.08	.48	.47	.39	2.0	12.2	11.9	9.9
807	6.4 x 22.2	0.250 x 0.875	.06	.38	.36	.25	1.5	9.7	9.1	6.4
808	6.4 x 25.4	0.250 x 1.000	.06	.44	.43	.31	1.5	11.2	10.9	7.9

Source: American Society of Mechanical Engineers

Table 15 Sheet Metal Gages and Thicknesses

North American Gages								European Gages			
Ferrous Metals, Such as Galvanized Steel, Tin Plate				Galvanized Steel, Tin Plate, Copper, Strip Steel and Steel, Copper, and Aluminum Tubes				Nonferrous Metals, Such as Copper, Brass, Aluminum		Nonferrous	
U.S. Standard (USS)		U.S. Standard (Revised)		Birmingham (BWG)		New Birmingham (BG)		Browne And Sharpe (B & S)		Imperial Standard (SWG)	
Gage	In.	Gage	In.	Gage	In.	Gage	In.	Gage	In.	Gage	In.
		3	.239					3	.229		
4	.234	4	.224	4	.238	4	.250	4	.204	4	.232
5	.219	5	.209	5	.220	5	.223	5	.182	5	.212
6	.203	6	.194	6	.203	6	.198	6	.162	6	.192
7	.188	7	.179	7	.180	7	.176	7	.144	7	.176
8	.172	8	.164	8	.165	8	.157	8	.129	8	.160
9	.156	9	.149	9	.148	9	.140	9	.114	9	.144
10	.141	10	.135	10	.134	10	.125	10	.102	10	.128
11	.125	11	.120	11	.120	11	.111	11	.091	11	.116
12	.109	12	.105	12	.109	12	.099	12	.081	12	.104
13	.094	13	.090	13	.095	13	.088	13	.072	13	.092
14	.078	14	.075	14	.083	14	.079	14	.064	14	.080
15	.070	15	.067	15	.072	15	.070	15	.057	15	.072
16	.063	16	.060	16	.065	16	.063	16	.051	16	.064
17	.056	17	.054	17	.058	17	.056	17	.045	17	.056
18	.050	18	.048	18	.049	18	.050	18	.040	18	.048
19	.044	19	.042	19	.042	19	.044	19	.036	19	.040
20	.038	20	.036	20	.035	20	.039	20	.032	20	.036
21	.034	21	.033	21	.032	21	.035	21	.029	21	.032
22	.031	22	.030	22	.028	22	.031	22	.025	22	.028
23	.028	23	.027	23	.025	23	.028	23	.023	23	.024
24	.025	24	.024	24	.022	24	.025	24	.020	24	.022
25	.022	25	.021	25	.020	25	.022	25	.018	25	.020
26	.019	26	.018	26	.018	26	.020	26	.016	26	.018
27	.017	27	.016	27	.016	27	.017	27	.014	27	.016
28	.016	28	.015	28	.014	28	.016	28	.013	28	.015
29	.014	29	.014	29	.013	29	.014	29	.011	29	.014
30	.012	30	.012	30	.012	30	.012	30	.010	30	.012
31	.011	31	.011	31	.010	31	.011	31	.009		
32	.010	32	.010	32	.009			32	.008	32	.011
33	.009	33	.009	33	.008	33	.009	33	.007	33	.010
34	.008	34	.008	34	.007	34	.008	34	.006	34	.009
				35	.005	35	.007			35	.008
36	.007	36	.007	36	.004	36	.006	36	.005		

Source: American Society of Mechanical Engineers

Table 16 Metric Conversion Tables

Quantity	Metric Unit	Symbol	Metric to Inch-Pound Unit	Inch-Pound to Metric Unit
Length	millimeter	mm	1 mm = 0.0394 in.	1 in. = 25.4 mm
	centimeter	cm	1 cm = 0.394 in.	1 ft. = 30.5 cm
	meter	m	1 m = 39.37 in. = 3.28 ft	1 yd. = 0.914 m = 914 mm
	kilometer	km	1 km = 0.62 mile	1 mile = 1.61 km
Area	square millimeter	mm²	1 mm² = 0.001 55 sq. in.	1 sq. in. = 6 452 mm²
	square centimeter	cm²	1 cm² = 0.155 sq. in.	1 sq. ft. = 0.093 m²
	square meter	m²	1 m² = 10.8 sq. ft. = 1.2 sq. yd.	1 sq. yd. = 0.836 m²
Mass	milligram	mg	1 g = 0.035 oz.	1 oz. = 28.3 g
	gram	g	1 kg = 2.205 lb.	1 lb. = 0.454 kg
	kilogram	kg	1 tonne = 1.102 tons	1 ton = 907.2 kg
	tonne	t		= 0.907 tonnes
Volume	cubic centimeter	cm³	1 mm³ = 0.000 061 cu. in.	1 fl. oz. = 28.4 cm³
	cubic meter	m³	1 cm³ = 0.061 cu. in.	1 cu. in. = 16.387 cm³
	milliliter	m	1 m³ = 35.3 cu. ft. = 1.308 cu. yd.	1 cu. ft. = 0.028 m³
			1 mℓ = 0.035 fl. oz.	1 cu. yd. = 0.756 m³
Capacity	liter	L	U.S. Measure 1 pt. = 0.473 L 1 pt. = 0.946 L 1 gal. = 3.785 L Imperial Measure 1 pt. = 0.568 L 1 qt. = 1.137 L 1 gal. = 4.546 L	U.S. Measure 1 L = 2.113 pt. = 1.057 qt. = 0.264 gal. Imperial Measure 1 L = 1.76 pt. = 0.88 qt. = 0.22 gal.
Temperature	Celsius degree	°C	$°C = \frac{5}{9}(°F\text{-}32)$	$°F = \frac{9}{5} \times °C + 32$
Force	newton	N	1 N = 0.225 lb (f)	1 lb (f) = 4.45N
	kilonewton	kN	1 kN = 0.225 kip (f) = 0.112 ton (f)	= 0.004 448 kN
Energy/Work	joule	J	1 J = 0.737 ft ∘ lb	1 ft ∘ lb = 1.355 J
	kilojoule	kJ	1 J = 0.948 Btu	1 Btu = 1.055 J
	megajoule	MJ	1 MJ = 0.278 kWh	1 kWh = 3.6 MJ
Power	kilowatt	kW	1 kW = 1.34 hp	1 hp (550 ft ∘ lb/s) = 0.746 kW
			1 W = 0.0226 ft ∘ lb/min.	1 ft ∘ lb/min = 44.2537 W
Pressure	kilopascal	kPa	1 kPa = 0.145 psi = 20.885 psf = 0.01 ton-force per sq. ft.	1 psi = 6.895 kPa 1 lb-force/sq. ft. = 47.88 Pa 1 ton-force/sq. ft. = 95.76 kPa
	*kilogram per square centimeter	kg/cm²	1 kg/cm² = 13,780 psi	
Torque	newton meter	N ∘ m	1 N ∘ m = 0.74 lb ∘ ft	1 lb ∘ ft = 1.36 N ∘ m
	*kilogram meter	kg/m	1 kg/m = 7.24 lb ∘ ft	1 lb ∘ ft = 0.14 kg/m
	*kilogram per centimeter	kg/cm	1 kg/cm = 0.86 lb ∘ in	1 lb ∘ in = 1.2 kg/cm
Speed/Velocity	meters per second	m/s	1 m/s = 3.28 ft/s	1 ft/s = 0.305 m/s
	kilometers per hour	km/h	1 km/h = 0.62 mph	1 mph = 1.61 km/h

*Not SI units, but included here because they are employed on some of the gages and indicators currently in use in industry.

INDEX

Note: The letters 't' and 'f' represents 'tables' and 'figures' respectively.